# Flat Level Set Regularity of *p*-Laplace Phase Transitions

# Memoirs
of the
American Mathematical Society

Number 858

# Flat Level Set Regularity of *p*-Laplace Phase Transitions

Enrico Valdinoci
Berardino Sciunzi
Vasile Ovidiu Savin

July 2006 • Volume 182 • Number 858 (second of 4 numbers) • ISSN 0065-9266

**American Mathematical Society**
Providence, Rhode Island

2000 *Mathematics Subject Classification.* Primary 35J70, 35B65.

---

**Library of Congress Cataloging-in-Publication Data**

Valdinoci, Enrico, 1974–
　Flat level set regularity of $p$-Laplace phase transitions / Enrico Valdinoci, Berardino Sciunzi, Vasile Ovidiu Savin.
　　p. cm. — (Memoirs of the American Mathematical Society, ISSN 0065-9266 ; no. 858)
　"Volume 182, number 858 (second of 4 numbers)."
　Includes bibliographical references.
　ISBN 0-8218-3910-1 (alk. paper)
　1. Geometry, Differential. 2. Laplacian operator. 3. Level set methods. I. Sciunzi, Berardino. II. Savin, Vasile Ovidiu, 1977– III. Title. IV. Series.
QA3.A57 no. 858
[QA641]
510 s—dc22
[515′.39]　　　　　　　　　　　　　　　　　　　　　　　　　　　　　　　　　　　　2006042822

---

# Memoirs of the American Mathematical Society

　This journal is devoted entirely to research in pure and applied mathematics.

**Subscription information.** The 2006 subscription begins with volume 179 and consists of six mailings, each containing one or more numbers. Subscription prices for 2006 are US$624 list, US$499 institutional member. A late charge of 10% of the subscription price will be imposed on orders received from nonmembers after January 1 of the subscription year. Subscribers outside the United States and India must pay a postage surcharge of US$31; subscribers in India must pay a postage surcharge of US$43. Expedited delivery to destinations in North America US$35; elsewhere US$130. Each number may be ordered separately; *please specify number* when ordering an individual number. For prices and titles of recently released numbers, see the New Publications sections of the *Notices of the American Mathematical Society*.

**Back number information.** For back issues see the *AMS Catalog of Publications*.

　Subscriptions and orders should be addressed to the American Mathematical Society, P. O. Box 845904, Boston, MA 02284-5904, USA. *All orders must be accompanied by payment.* Other correspondence should be addressed to 201 Charles Street, Providence, RI 02904-2294, USA.

**Copying and reprinting.** Individual readers of this publication, and nonprofit libraries acting for them, are permitted to make fair use of the material, such as to copy a chapter for use in teaching or research. Permission is granted to quote brief passages from this publication in reviews, provided the customary acknowledgment of the source is given.

　Republication, systematic copying, or multiple reproduction of any material in this publication is permitted only under license from the American Mathematical Society. Requests for such permission should be addressed to the Acquisitions Department, American Mathematical Society, 201 Charles Street, Providence, Rhode Island 02904-2294, USA. Requests can also be made by e-mail to `reprint-permission@ams.org`.

---

*Memoirs of the American Mathematical Society* is published bimonthly (each volume consisting usually of more than one number) by the American Mathematical Society at 201 Charles Street, Providence, RI 02904-2294, USA. Periodicals postage paid at Providence, RI. Postmaster: Send address changes to Memoirs, American Mathematical Society, 201 Charles Street, Providence, RI 02904-2294, USA.

　　　　© 2006 by the American Mathematical Society. All rights reserved.
　　　Copyright of this publication reverts to the public domain 28 years
　　　　　after publication. Contact the AMS for copyright status.
　This publication is indexed in *Science Citation Index*®, *SciSearch*®, *Research Alert*®,
　　*CompuMath Citation Index*®, *Current Contents*®/*Physical, Chemical & Earth Sciences*.
　　　　　　　Printed in the United States of America.

　　　∞ The paper used in this book is acid-free and falls within the guidelines
　　　　　　established to ensure permanence and durability.
　　　　　　Visit the AMS home page at `http://www.ams.org/`

　　　　　　10 9 8 7 6 5 4 3 2 1　　11 10 09 08 07 06

# Contents

| | | |
|---|---|---:|
| Chapter 1. | Introduction | 1 |
| Chapter 2. | Modifications of the potential and of one-dimensional solutions | 7 |
| Chapter 3. | Geometry of the touching points | 31 |
| Chapter 4. | Measure theoretic results | 43 |
| Chapter 5. | Estimates on the measure of the projection of the contact set | 47 |
| Chapter 6. | Proof of Theorem 1.1 | 61 |
| Chapter 7. | Proof of Theorem 1.2 | 69 |
| Chapter 8. | Proof of Theorem 1.3 | 75 |
| Chapter 9. | Proof of Theorem 1.4 | 79 |
| Appendix A. | Proof of the measure theoretic results | 83 |
| A.1. | Proof of Lemma 4.1 | 83 |
| A.2. | Proof of Lemma 4.2 | 88 |
| A.3. | Proof of Lemma 4.3 | 127 |
| Appendix B. | Summary of elementary lemmata | 135 |
| Bibliography | | 143 |

# Abstract

We prove a Harnack inequality for level sets of $p$-Laplace phase transition minimizers. In particular, if a level set is included in a flat cylinder, then, in the interior, it is included in a flatter one. The extension of a result conjectured by De Giorgi and recently proven by the third author for $p = 2$ follows.

---

Received by the editor January 14, 2005.
2000 *Mathematics Subject Classification.* 35J70, 35B65.
*Key words and phrases.* Ginzburg-Landau-Allen-Cahn phase transition models, De Giorgi conjecture, $p$-Laplacian operator, sliding methods, geometric and qualitative properties of solutions.

We thank Xavier Cabré for an interesting discussion. EV would like to thank also Alberto Farina for first mentioning the conjecture of De Giorgi in a pleasant meeting at the Erwin Schroedinger Institute (Vienna, 2002). EV and BS have been supported by MIUR *Variational Methods and Nonlinear Differential Equations*.

# CHAPTER 1

# Introduction

Given a domain $\Omega \subseteq \mathbb{R}^N$, we define the following functional on $W^{1,p}(\Omega)$:

$$\mathcal{F}_\Omega(u) = \int_\Omega \frac{|\nabla u(x)|^p}{p} + h_0(u(x))\,dx\,.$$

Here above and in the sequel, we suppose that $1 < p < \infty$ and that $h_0 \in C^0([-1,1]) \cap C^{1,1}((-1,1))$ can be extended to a function which is $C^1$ in a neighborhood of $[-1,1]$. We will also assume that, for some $0 < c < 1 < C$ and some $\theta^\star \in (0,1)$, we have

(1.1) $$\text{for any } \theta \in [0,1],\ c\,\theta^p \leq h_0(-1+\theta) \leq C\,\theta^p \text{ and}$$
$$c\,\theta^p \leq h_0(1-\theta) \leq C\,\theta^p,$$

(1.2) $$\text{for any } \theta \in [0,\theta^\star),\ h_0'(-1+\theta) \geq c\theta^{p-1} \text{ and}$$
$$h_0'(1-\theta) \leq -c\theta^{p-1}.$$

We also assume that $h_0'$ is monotone increasing in $(-1,-1+\theta^\star) \cup (1-\theta^\star,1)$. Quantities depending only on the constants above will be referred to as "universal constants". As a model example for a potential $h_0$ satisfying the conditions stated here above, one may consider

$$h_0(\zeta) := (1-\zeta^2)^p\,.$$

In the literature, $h_0$ is often referred to as a "double-well" potential, and its derivative as a "bi-stable nonlinearity".

In the light of (1.1) and (1.2), we have that, with no loss of generality, possibly reducing the size of $\theta^\star$, we may and do assume that

(1.3) $$\text{for any } \zeta \in [-1+\theta^\star, 1-\theta^\star],$$
$$h_0(\zeta) \geq \max\nolimits_{[-1,-1+\theta^\star] \cup [1-\theta^\star,1]} h_0,$$

Notice that, if $u \in W^{1,p}(\Omega)$, $|u| \leq 1$, is[1] critical for $\mathcal{F}_\Omega$, then $u$ satisfies in the weak sense the following singular/degenerate elliptic equation of $p$-Laplacian type:

(1.5) $$\Delta_p u(x) = h_0'(u(x)),$$

for any $x \in \Omega$. Here and in what follows, we make use of the standard notation

$$\Delta_p u := \text{div}\left(|\nabla u|^{p-2} \nabla u\right).$$

In particular, we will consider local minimizers for the functional above. We say that $u$ is a local minimizer for $\mathcal{F}$ in the domain $\Omega$ if

$$\mathcal{F}_\Omega(u) \leq \mathcal{F}_\Omega(u + \phi),$$

for any $\phi \in C_0^\infty(\Omega)$. In the literature, it is also customary to say that $u$ is a Class A minimizer for $\mathcal{F}$ if

$$\mathcal{F}_K(u) \leq \mathcal{F}_K(u + \phi),$$

for any compact set $K \subset \mathbb{R}^N$ and any $\phi \in C_0^\infty(K)$. That is, $u$ is a Class A minimizer if it is a local minimizer in any domain.

The functional $\mathcal{F}$ here above has been widely studied both for pure mathematical reasons and for physical applications. For instance, this functional is a model for interfaces appearing in physical problems when two phases (the phase "close to $+1$" and the one "close to $-1$") coexist. On one side, the "potential" $h_0$ tends to drive the minima of the functional towards the "pure states" $\pm 1$; on the other hand, the "kinetic term" $|\nabla u|^p$ prevents the system from sudden phase changes. The balance between these tendencies (or, in the physical language, the effect of the surface tension) leads interfaces of minimal solution to minimize area. The physical relevance of the interfaces and the mathematical interest arising from geometric measure theory thus motivated an extensive study of the transition layers, i.e., of the level sets of solutions of (1.5). We refer to [3], [7], [14], [23], [25], [26], [33], [27], [28], [29], [27], [28], [30] and [36] for more detailed discussions on the physical relevance of the above functional and for its relation with the theory of minimal surfaces.

The main result that will be proved in this paper is the following Harnack inequality for level sets of minimizers. Roughly speaking, such results says that, once one knows that the zero level set of a minimizer is trapped in a rectangle whose height is small enough, then, in a smaller neighborhood, it can be trapped in a rectangle with even smaller height. More precisely, we have the following result:

THEOREM 1.1. *Let $l > 0$, $\theta > 0$. Let $u$ be a local minimizer for $\mathcal{F}$ in*

$$\left\{(x', x_N) \in \mathbb{R}^{N-1} \times \mathbb{R} \,\middle|\, |x'| < l,\, |x_N| < l\right\}.$$

*Assume that $u(0) = 0$ and that*

$$\{u = 0\} \subseteq \{|x_N| < \theta\}.$$

---

[1] Many results of this paper are obtained without assuming $|u| < 1$, but assuming only $|u| \leq 1$. For future use, however, we recall that the condition $|u| < 1$ is fulfilled by any solution $u$ such that $|u| \leq 1$ with $|u|$ not identically equal to $\pm 1$, under suitable assumptions on $h_0$. This holds, for instance, if we suppose that

(1.4) $$h_0'(-1 + \theta) \leq c'\theta^{p-1} \quad \text{and} \quad h_0'(1 - \theta) \geq -c'\theta^{p-1},$$

for any $\theta \in [0, \theta^*)$. For the proof of this observation, see, e.g., footnote 7 in [30]. The case $p = 2$ was also dealt with in Theorem 1.1 of [17].

*Then, there exists a universal constant $c \in (0,1)$ so that, for any $\theta_0 > 0$ there exists $\varepsilon_0(\theta_0) > 0$ such that, if*

$$\frac{\theta}{l} \leq \varepsilon_0(\theta_0) \quad \text{and} \quad \theta \geq \theta_0,$$

*then*

$$\{u = 0\} \cap \{|x'| < cl\} \subseteq \{|x_N| < (1-c)\theta\}.$$

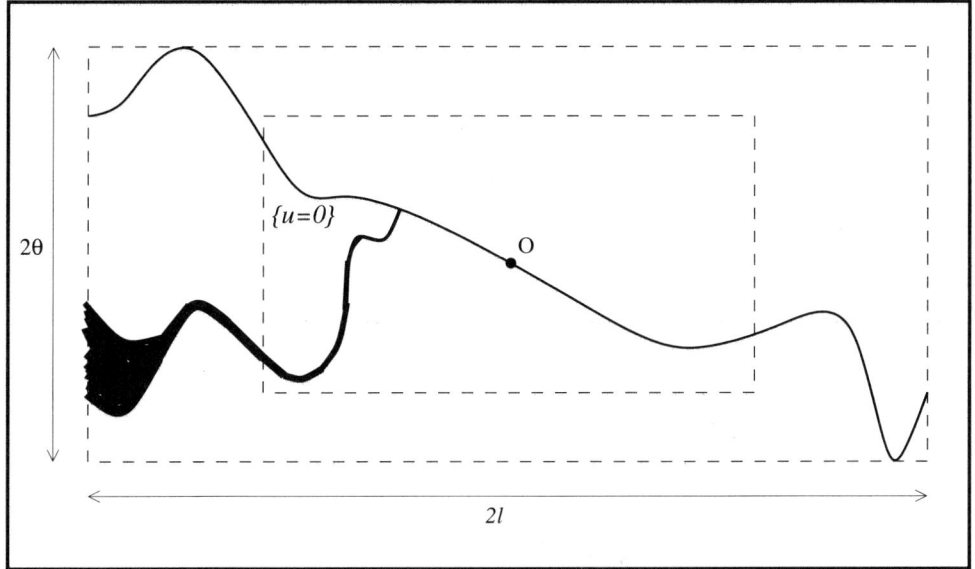

**The Harnack-type result of Theorem 1.1**

Theorem 1.1 is an extension of a similar result obtained in [**31**] for $p = 2$. Also, some results from [**7**], [**28**] and [**30**] will be needed in the course of the proof.

The proof of Theorem 1.1 is quite long, both because we will need some fine analysis on the measure estimates of the touching points between $u$ and some appropriate barriers, and because some delicate details and technical points will appear in the course of the proof. Very roughly, we can say that the final target of the proof consists in deducing a measure estimates on the above mentioned contact points, which, in case the statement of Theorem 1.1 were false, would contradict the minimality of $u$. Such estimates will be obtained by sliding suitable barriers, constructed via the one-dimensional heteroclinic solution.

The ideas of such proof become more transparent in the easier case of a uniformly elliptic equation involving principal curvatures (of the type $\sum_{i=1}^{N} a_i \kappa_i = 0$): see [**32**].

Following the ideas of [**31**], several results may be deduced from Theorem 1.1. First, we deduce the following "flatness improvement" result, stating that, once a level set is trapped inside a flat cylinder, then, possibly changing coordinates, it is trapped in an even flatter cylinder in the interior. More precisely, we have:

# 1. INTRODUCTION

THEOREM 1.2. *Let $l > 0$, $\theta > 0$. Let $u$ be a local minimizer for $\mathcal{F}$ in*
$$\left\{(x', x_N) \in \mathbb{R}^{N-1} \times \mathbb{R} \,\Big|\, |x'| < l,\, |x_N| < l\right\}.$$
*Assume that $u(0) = 0$ and that*
$$\{u = 0\} \subseteq \{|x_N| < \theta\}.$$
*Then, there exist universal constants $\eta_1, \eta_2 > 0$, with $0 < \eta_1 < \eta_2 < 1$, such that, for any $\theta_0 > 0$, there exists $\varepsilon_1(\theta_0) > 0$ such that, if*
$$\frac{\theta}{l} \leq \varepsilon_1(\theta_0) \qquad \text{and} \qquad \theta \geq \theta_0,$$
*then*
$$\{u = 0\} \cap \left(\{|\pi_\xi x| < \eta_2 l\} \times \{|(x \cdot \xi)| < \eta_2 l\}\right) \subseteq$$
$$\subseteq \left(\{|\pi_\xi x| < \eta_2 l\} \times \{|(x \cdot \xi)| < \eta_1 \theta\}\right)$$
*for some unit vector $\xi$.*

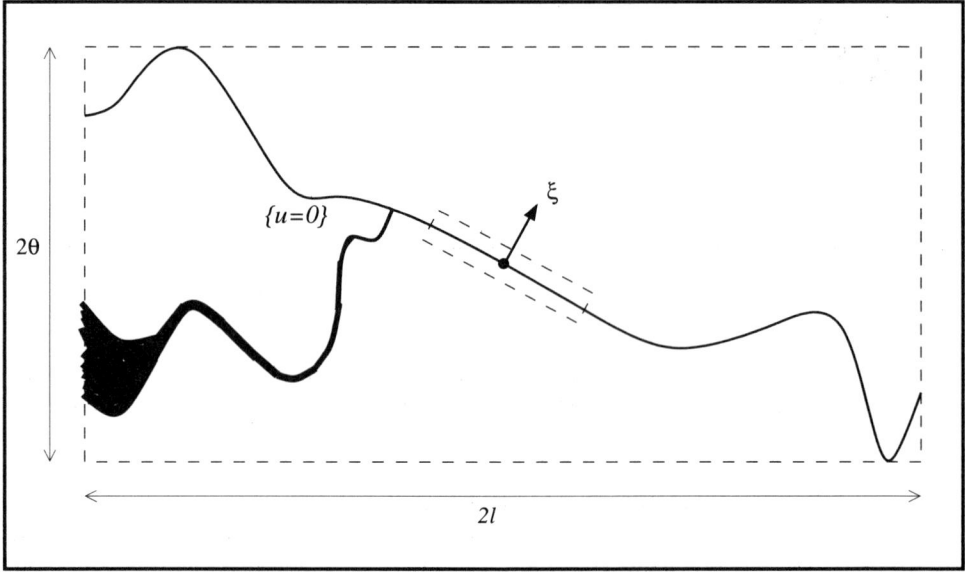

**The flatness improvement of Theorem 1.2**

Several ideas related with Theorem 1.2 have been extensively used by De Giorgi and his school in the minimal surface setting (e.g., for proving smoothness and analytic regularity): see, for instance, chapters 6–8 in [**20**].

The extension of a result conjectured by De Giorgi in [**14**] for $p = 2$ also follows, namely we have the following two flatness results:

THEOREM 1.3. *Let $N \leq 7$. Then, level sets of Class A minimizers of $\mathcal{F}$ are hyperplanes.*

THEOREM 1.4. *Let $u \in W^{1,p}_{\text{loc}}(\mathbb{R}^N)$ be a solution of (1.5). Let $h_0$ fulfill the assumptions on page 1 and[2] (1.4). Assume that $|u| \leq 1$, $\partial_N u > 0$ and*

$$\lim_{x_N \to \pm\infty} u(\cdot, x_N) = \pm 1\,.$$

*Assume also that either $N \leq 8$ or that $\{u = 0\}$ has at most linear growth at infinity. Then, level sets of $u$ are hyperplanes.*

Results of these type have been proved in [**21**] for $p = N = 2$ and [**2**] for $p = 2$ and $N = 3$ (and actually for any nonlinearity, see [**1**]). Extensions of the results in [**21**] and [**2**] to $p$-Laplace equations have been considered in [**11**]. See also [**5, 13, 12, 16, 4, 6, 22**] for related results. Results analogous to Theorems 1.3 and 1.4 for $p = 2$ have been recently given by the third author in [**31**]. In §9 here below, we will see that Theorem 1.4 is a consequence of the fact that monotone solutions of (1.5) are minimizers (see, e.g., [**24**] or [**31**]) under a $\pm 1$-limit assumption.

This paper is organized in the following way. In §2 we construct the barriers to be used in the course of the proofs of the main results. Roughly speaking, such barriers are obtained by modifying the heteroclinic one-dimensional solutions given by the potential $h_0$ and by taking flat or rotational extensions. The study of the touching points between these barriers and our solution occupies §3. Particular emphasis is given to the measure of the projection of the set of "first time" touches. In §4, some covering lemmas are presented, whose proof has been deferred to the Appendix. The results of §3 and §4 are then used in §5 to obtain an estimate on the projection of the touching points between an appropriate barrier and our solution. The proofs of the main results occupy §6—§9. The Appendix contains the proof of the covering lemmas and some elementary ancillary results.

---

[2]In particular, the result of Theorem 1.4 holds for $h_0(\zeta) := (1 - \zeta^2)^p$.

# CHAPTER 2

# Modifications of the potential and of one-dimensional solutions

We now construct some barriers, which will be of use in the proof of the main results. Such barriers will be obtained by appropriate modifications on the potential $h_0$, which induce corresponding modifications on one-dimensional solutions.

Here and below, we fix $\overline{C}_0 > 0$, to be conveniently chosen in the following (actually, during the proof of Proposition 2.13 here below). We will also fix $R$, to be assumed suitably large (with respect to $\overline{C}_0$ and some universal constants). The first function needed in our construction is the following modification of the potential $h_0$ in the interval $[-3/4, 3/4]$:

DEFINITION 2.1. Fix $|s_0| \leq 1/4$. For any $|s| \leq 3/4$, we define[1]

$$(2.1) \qquad \varphi_{s_0, R}(s) = \varphi(s) := \frac{h_0(s) R^p}{\left[R - \overline{C}_0(s - s_0)(\frac{p}{p-1} h_0(s))^{\frac{1}{p}}\right]^p}.$$

Note that, by construction,

$$(2.2) \qquad \frac{1}{\left(\frac{p}{p-1}\varphi(s)\right)^{\frac{1}{p}}} = \frac{1}{\left(\frac{p}{p-1} h_0(s)\right)^{\frac{1}{p}}} - \frac{\overline{C}_0}{R}(s - s_0).$$

Roughly speaking, for large $R$, $\varphi$ is close to $h_0$: this is the reason for which we consider $\varphi$ as a modified potential in $[-3/4, 3/4]$. We now consider some properties enjoyed by $\varphi$. First of all, we estimate $\varphi$ in terms of $h_0$ in $[-3/4, -1/2] \cup [1/2, 3/4]$:

LEMMA 2.2. *The following inequalities hold:*

$$(2.3) \qquad \varphi(s) < h_0(s) - \frac{2\widehat{C}_0}{R} \quad \text{if} \quad s \in \left[-\frac{3}{4}, -\frac{1}{2}\right]$$

*and*

$$(2.4) \qquad \varphi(s) > h_0(s) + \frac{2\widehat{C}_0}{R} \quad \text{if} \quad s \in \left[\frac{1}{2}, \frac{3}{4}\right],$$

*provided that $R$ and $\overline{C}_0/\widehat{C}_0$ are suitably large. Also, $\widehat{C}_0$ may be taken large if so is $\overline{C}_0$.*

PROOF. To prove (2.3), note that for $s \in [-\frac{3}{4}, -\frac{1}{2}]$ we have $\overline{C}_0(s - s_0) \in [-\overline{C}_0, -\frac{\overline{C}_0}{4}]$. Also, from (1.1), there exists $k > 0$ such that

$$(2.5) \qquad 0 < k \leq \inf_{\sigma \in [-3/4, -1/2] \cup [1/2, 3/4]} \left(\frac{p}{p-1} h_0(\sigma)\right)^{\frac{1}{p}}.$$

---

[1] Notice that $R - \overline{C}_0(s - s_0)(\frac{p}{p-1} h_0(s))^{\frac{1}{p}} > 0$ for any $|s| \leq 3/4$ and $|s_0| \leq 1/4$, if $R$ is large enough, thus the definition of $\varphi$ is well posed.

Therefore,

$$\varphi(s) = \frac{h_0(s)R^p}{\left[R - \overline{C}_0(s-s_0)(\frac{p}{p-1}h_0(s))^{\frac{1}{p}}\right]^p} \leq$$

$$\leq \frac{h_0(s)R^p}{(R + \frac{\overline{C}_0}{4}k)^p} =$$

$$= h_0(s) + \frac{h_0(s)R^p - h_0(s)(R + \frac{\overline{C}_0}{4}k)^p}{(R + \frac{\overline{C}_0}{4}k)^p} \leq$$

$$\leq h_0(s) - \frac{\text{const}}{R} \cdot \frac{(R + \text{const}\,\overline{C}_0)^p - R^p}{R^{p-1}}.$$

Using the fact that

$$\lim_{x \to +\infty} \frac{(x+a)^p - x^p}{x^{p-1}} = pa,$$

and taking $R$ suitably large, we get

$$\varphi(s) \leq h_0(s) - \frac{\text{const}\,\overline{C}_0}{R} \leq h_0(s) - \frac{2\widehat{C}_0}{R},$$

for $\overline{C}_0 > \text{const}\,\widehat{C}_0$, proving (2.3). Let us now prove (2.4). Recalling (2.5) and arguing as above, for $s \in [1/2, 3/4]$, we get that $\overline{C}_0(s-s_0) \in [\frac{\overline{C}_0}{4}, \overline{C}_0]$ and so

$$\varphi(s) = \frac{h_0(s)R^p}{\left[R - \overline{C}_0(s-s_0)(\frac{p}{p-1}h_0(s))^{\frac{1}{p}}\right]^p} \geq$$

$$\geq \frac{h_0(s)R^p}{(R - \frac{\overline{C}_0}{4}k)^p} \geq$$

$$\geq h_0(s) + \frac{\text{const}}{R} \cdot \frac{R^p - (R - \text{const}\,\overline{C}_0)^p}{R^{p-1}}.$$

Using the fact that

$$\lim_{x \to +\infty} \frac{x^p - (x-a)^p}{x^{p-1}} = pa,$$

we thus gather that, for $R$ large,

$$\varphi(s) \geq h_0(s) + \frac{\text{const}\,\overline{C}_0}{R} \geq h_0(s) + \frac{2\widehat{C}_0}{R},$$

for $\overline{C}_0 > \text{const}\,\widehat{C}_0$, proving (2.4). $\square$

Now, for $R$ large enough, we define $s_R \in (-1, \theta^*)$ as the point such that

(2.6) $$h_0(s_R) = \frac{1}{R}.$$

From (1.1), we have that

(2.7) $$c(1 + s_R) \leq \frac{1}{R^{1/p}} \leq C(1 + s_R).$$

For further estimates, in the next two lemmas, we now point out some elementary bounds for $s_R$:

## 2. MODIFICATIONS OF THE POTENTIAL AND OF ONE-DIMENSIONAL SOLUTIONS

LEMMA 2.3. *Let $C$ a positive universal constant. Then, for $R$ large (with respect to $C$), we have that*

$$(2.8) \qquad (1+\xi)^p - (1+s_R)^p - \frac{C}{R}(\xi - s_R) > 0 \qquad \text{if} \quad s_R < \xi < 0$$

*and*

$$(2.9) \qquad \int_{s_R}^0 \frac{d\xi}{\left((1+\xi)^p - (1+s_R)^p - \frac{C}{R}(\xi - s_R)\right)^{1/p}} \leq \widetilde{C} \log R,$$

*for a suitable constant $\widetilde{C} > 0$.*

PROOF. For $\xi \in [s_R, 0]$, let

$$g(\xi) := (1+\xi)^p - \frac{C}{R}(1+\xi).$$

Then, by (2.7),

$$\begin{aligned} g'(\xi) &= p(1+\xi)^{p-1} - \frac{C}{R} \geq \\ &\geq p(1+s_R)^{p-1} - \frac{C}{R} \geq \\ &\geq \frac{\text{const}}{R^{(p-1)/p}} - \frac{C}{R} > 0, \end{aligned}$$

if $R$ is large enough, thence $g(\xi) > g(s_R)$, proving (2.8).

In order to estimate the integral in (2.9), we introduce the notation $b = 1 + s_R$. Since, for $R$ large, $s_R$ is near $-1$, we may assume $\frac{1}{b} > 2$. We use the substitution

$$\tau = \frac{1+\xi}{b}$$

obtaining a bound for the above integral given by

$$\int_1^{1/b} \frac{d\tau}{\left(\tau^p - 1 - \frac{C}{Rb^p}b(\tau-1)\right)^{1/p}} \leq \int_1^{1/b} \frac{d\tau}{\left(\tau^p - 1 - C'b(\tau-1)\right)^{1/p}} \leq$$

$$\leq \int_1^2 \frac{d\tau}{\left(\tau^p - 1 - C'b(\tau-1)\right)^{1/p}} + \int_2^{1/b} \frac{d\tau}{\left(\tau^p - 1 - C'b(\tau-1)\right)^{1/p}}.$$

where we used the fact that $b^p \approx \frac{1}{R}$ (see (2.7)).

Noticing that

$$\tau^p - 1 \geq \frac{2^p - 1}{2^p} \tau^p$$

if $\tau \geq 2$, and that

$$\tau^p - 1 \geq \tau - 1$$

if $\tau \geq 1$, we bound the quantity here above by

$$\text{const} \left( \int_1^2 \frac{d\tau}{((1-C'b)(\tau-1))^{1/p}} + \int_2^{1/b} \frac{d\tau}{\left(\tau^p\left(1 - \frac{C'b(\tau-1)}{\tau^p}\right)\right)^{\frac{1}{p}}} \right).$$

Now, if $R$ is large, then $b$ is small, and therefore we may assume that $(1-C'b)^{\frac{1}{p}} > \frac{1}{2}$. Thus,

$$\left(1 - \frac{C'b(\tau-1)}{\tau^p}\right)^{\frac{1}{p}} \geq \left(1 - \frac{C'b}{\tau^{p-1}}\right)^{\frac{1}{p}} \geq$$

$$\geq (1-C'b)^{\frac{1}{p}} > \frac{1}{2}.$$

This yields

$$\int_{s_R}^{0} \frac{d\xi}{\left((1+\xi)^p - (1+s_R)^p - \frac{C}{R}(\xi - s_R)\right)^{1/p}} \leq \text{const}\left(1 + \log\frac{1}{2b}\right),$$

which proves (2.9). □

Let us now estimate how $s_R$ varies as a function of $R$:

LEMMA 2.4. *There exists a suitable universal constant $C > 0$ so that*

$$-\frac{C}{R^{(p+1)/p}} \leq \partial_R s_R < 0.$$

PROOF. Differentiating (2.6),

$$-\frac{1}{R^2} = h_0'(s_R)\,\partial_R s_R,$$

thus $\partial_R s_R < 0$ thanks to (1.2), and so, by (2.7),

$$\frac{1}{R^2} = h_0'(s_R)\,|\partial_R s_R| \geq$$

$$\geq \text{const}\,(1+s_R)^{p-1}\,|\partial_R s_R| \geq$$

$$\geq \frac{\text{const}}{R^{(p-1)/p}}\,|\partial_R s_R|.$$

□

We now define a modification of the potential $h_0$ in the whole interval $[-1,1]$ in the following way:

DEFINITION 2.5. We define $h_{s_0,R} : [s_R, 1] \to \mathbb{R}$ by

$$h_{s_0,R}(s) := \begin{cases} h_0(s) - h_0(s_R) - \frac{\widehat{C_0}}{R}(s - s_R) \\ \quad \text{if} \quad s_R \leq s \leq -\frac{1}{2} \\ \varphi(s) \\ \quad \text{if} \quad -\frac{1}{2} < s < \frac{1}{2} \\ h_0(s) + h_0(s_R) + \frac{\widehat{C_0}}{R}(1-s) \\ \quad \text{if} \quad \frac{1}{2} \leq s \leq 1. \end{cases}$$

Notice that $h_{s_0,R}$ may be discontinuous at $s = \pm 1/2$. Let us now point out some easy properties enjoyed by the above potential:

## 2. MODIFICATIONS OF THE POTENTIAL AND OF ONE-DIMENSIONAL SOLUTIONS

LEMMA 2.6. *The following inequalities hold. If $s \in [-3/4, -1/2]$, then*

$$h_{s_0, R}(s) > h_0(s) - \frac{2\widehat{C}_0}{R} > \varphi(s).$$

*If $s \in [1/2, 3/4]$, then*

$$h_{s_0, R}(s) < h_0(s) + \frac{2\widehat{C}_0}{R} < \varphi(s).$$

PROOF. Let $s \in [-3/4, -1/2]$. If $\widehat{C}_0$ is taken suitably large, then, recalling Lemma 2.2, we have that

$$h_{s_0, R}(s) = h_0(s) - \frac{1}{R} - \frac{\widehat{C}_0}{R}(s - s_R) \geq$$
$$\geq h_0(s) - \frac{2\widehat{C}_0}{R} > \varphi(s)$$

where we have used the fact that $s - s_R \leq 1$.

In the same way, if $s \in [1/2, 3/4]$, taking $\widehat{C}_0$ suitably large, and using Lemma 2.2, we gather that

$$h_{s_0, R} < h_0(s) + \frac{1}{R} + \frac{\widehat{C}_0}{2R} <$$
$$< h_0(s) + \frac{2\widehat{C}_0}{R} < \varphi(s).$$

□

LEMMA 2.7. *Let $R_1 \leq R_2$ be suitably large. Then,*[2]

(2.10) $\qquad h_{s_0, R_1}(s) \leq h_{s_0, R_2}(s) \qquad \text{if} \quad s_{R_1} \leq s \leq s_0$

*and*

(2.11) $\qquad h_{s_0, R_1}(s) \geq h_{s_0, R_2}(s) \qquad \text{if} \quad s_0 \leq s \leq 1.$

PROOF. Let us prove (2.10). For this purpose, let $s \leq s_0$. Two cases are possible: either $s > -1/2$ or $s \leq -1/2$. Let us first deal with the first case. Notice that, by (2.1), $\varphi_{s_0, R}$ is increasing in $R$, since $-1/2 < s \leq s_0 \leq 1/4 < 1/2$, thus, from Definition 2.5 we gather that

$$h_{s_0, R_1}(s) = \varphi_{s_0, R_1}(s) \leq$$
$$\leq \varphi_{s_0, R_2}(s) = h_{s_0, R_2}(s),$$

which proves (2.10) if $s > -1/2$.

Let us now deal with the case $s \leq -1/2$. Fixed $R_1$ large and $s \in [s_R, -1/2]$, let us define

$$g(R) := \frac{1}{R} + \frac{\widehat{C}_0}{R}(s - s_R).$$

---

[2] Notice that, for $R_1 \leq R_2$ suitably large, using (1.2), (2.6) and (2.7), one has

$$h_0(s_{R_1}) = \frac{1}{R_1} \geq \frac{1}{R_2} = h_0(s_{R_2}),$$

so that $s_{R_1} \geq s_{R_2}$.

By means of Lemma 2.4,
$$g'(R) \leq -\frac{1}{R^2} + \frac{\operatorname{const} \widehat{C}_0}{R^{2+1/p}} < 0,$$
thence
$$\frac{1}{R_2} + \frac{\widehat{C}_0}{R_2}(s - s_{R_2}) = g(R_2) \leq$$
$$\leq g(R_1) = \frac{1}{R_1} + \frac{\widehat{C}_0}{R_1}(s - s_{R_1}).$$
Therefore, if $s \leq -1/2$, from Definition 2.5,
$$h_{s_0, R_1}(s) = h_0(s) - \frac{1}{R_1} - \frac{\widehat{C}_0}{R_1}(s - s_{R_1}) \leq$$
$$\leq h_0(s) - \frac{1}{R_2} - \frac{\widehat{C}_0}{R_2}(s - s_{R_2}) =$$
$$= h_{s_0, R_2}(s),$$
thus proving (2.10).

Having completed the proof of (2.10), we now deal with the proof of (2.11). Two cases are possible: either $s < 1/2$ or $s \geq 1/2$. Let us first deal with the first case. Notice that, by (2.1) and the fact that $s_0 \leq s < 1/2$, $\varphi_{s_0, R}$ is decreasing in $R$. Hence, from Definition 2.5,
$$h_{s_0, R_1}(s) = \varphi_{s_0, R_1}(s) \geq$$
$$\geq \varphi_{s_0, R_2}(s) = h_{s_0, R_2}(s),$$
which proves (2.11) if $s < 1/2$.

Let us now deal with the case $s \geq 1/2$. In this case, by Definition 2.5,
$$h_{s_0, R_1}(s) - h_{s_0, R_2}(s) = \frac{1}{R_1} - \frac{1}{R_2} + \widehat{C}_0(1-s)\left(\frac{1}{R_1} - \frac{1}{R_2}\right) \geq 0,$$
proving (2.11) for $s \geq 1/2$ and thus completing the proof of Lemma 2.7. □

Let now
$$(2.12) \qquad H_0(s) := \int_0^s \frac{(p-1)^{\frac{1}{p}}}{(p\,h_0(\zeta))^{\frac{1}{p}}}\,d\zeta, \qquad \text{for any } s \in (-1, 1).$$

Notice that the inverse of $H_0$ is a "one-dimensional" solution of (1.5). Indeed, if $g_0 := H_0^{-1}$, by Lemma B.3, we obtain that
$$\Delta_p g_0 = (|g'|^{p-2} g')' =$$
$$= (p-1)|g'|^{p-2} g'' =$$
$$= h_0'(g_0).$$

We would like now to compare $g_0$ with all other solutions.

To this aim, using Definition 2.5 we now introduce suitable modifications of $H_0$ (and thus of $g_0$), which will be used in the course of the proof:

## 2. MODIFICATIONS OF THE POTENTIAL AND OF ONE-DIMENSIONAL SOLUTIONS

DEFINITION 2.8. We define, for $s \in [s_R, 1]$,

$$H_{s_0,R}(s) := H_0(s_0) + \int_{s_0}^{s} \frac{(p-1)^{\frac{1}{p}}}{(p\, h_{s_0,R}(\zeta))^{\frac{1}{p}}}\, d\zeta\,.$$

We notice that, by Lemmas B.1 and 2.3, we get

(2.13) $$h_{s_0,R} > 0,$$

thence $H_{s_0,R}$ is well defined.

REMARK 2.9. By Definition 2.8, exploiting Lemma 2.7, it follows that, if $R_1 \leq R_2$,

(2.14) $$H_{s_0,R_1}(s) \leq H_{s_0,R_2}(s) \quad \text{if} \quad s_{R_1} \leq s \leq 1\,.$$

Let us now analyze some properties of the above defined quantity:

LEMMA 2.10. *Assuming $R$ suitably large, there exists a positive constant $C_1$ so that the following inequalities hold:*

(2.15) $$H_{s_0,R}(s_R) \geq -C_1 \log R,$$

(2.16) $$H_{s_0,R}(1) \leq C_1 \log R,$$

(2.17) $$\frac{d}{ds}(H_{s_0,R}(s)) > 0, \quad \forall s \in (s_R, 1)\,.$$

PROOF. Note that, if $R$ is suitably big, Definition 2.1 implies that

(2.18) $$\inf_{|s|\leq 1/2} \varphi(s) > 0\,.$$

Hence, recalling Lemma B.1 and (2.7), we have that

$$\begin{aligned} H_{s_0,R}(s_R) &\geq H_0(s_0) - \int_{s_R}^{-\frac{1}{2}} \frac{\text{const}\, d\zeta}{\left((1+\zeta)^p - (1+s_R)^p - \frac{C}{R}(\zeta - s_R)\right)^{1/p}} - \\ &\quad - \int_{-\frac{1}{2}}^{s_0} \frac{(p-1)^{\frac{1}{p}}}{(p\,\varphi(\zeta))^{\frac{1}{p}}}\, d\zeta \geq \\ &\geq -\text{const} - \int_{s_R}^{0} \frac{\text{const}\, d\zeta}{\left((1+\zeta)^p - (1+s_R)^p - \frac{C}{R}(\zeta - s_R)\right)^{1/p}} \geq \\ &\geq -C_1 \log R, \end{aligned}$$

for a suitable $C_1$, where we used Lemma 2.3 to estimate the integral above. This proves (2.15).

We now prove (2.16). By Definition 2.5, (1.1), (2.18) and (2.6), we have that

$$\begin{aligned} H_{s_0,R}(1) &\leq H_0(s_0) + \int_{s_0}^{\frac{1}{2}} \frac{(p-1)^{\frac{1}{p}}}{(p\,\varphi(\zeta))^{\frac{1}{p}}}\, d\zeta + \int_{\frac{1}{2}}^{1} \frac{\text{const}\, d\zeta}{\left((1-\zeta)^p + \frac{1}{R}\right)^{1/p}} \leq \\ &\leq \text{const} + \int_{1/2}^{1} \frac{\text{const}\, d\zeta}{(1-\zeta) + \frac{1}{R^{1/p}}} \leq \\ &\leq \text{const}\left(1 + \log \frac{R^{1/p} + 2}{2}\right) \leq \\ &\leq C_1 \log R, \end{aligned}$$

proving (2.16). Finally, (2.17) follows from Definition 2.8 and (2.13). □

Using (2.17), we may now give the following definition:

DEFINITION 2.11. We define $g_{s_0,R}(t) : (-\infty, H_{s_0,R}(1)] \to \mathbb{R}$ by

$$g_{s_0,R}(t) := \begin{cases} s_R & \text{if } t \leq H_{s_0,R}(s_R), \\ H_{s_0,R}^{-1}(t) & \text{if } H_{s_0,R}(s_R) < t < H_{s_0,R}(1). \end{cases}$$

We now state some notation. Given $X \in \mathbb{R}^{N+1}$ we define $x \in \mathbb{R}^N$ and $x_{N+1} \in \mathbb{R}$ in such a way

$$X = (x, x_{N+1}) \in \mathbb{R}^N \times \mathbb{R}.$$

Also, we will often denote

$$x = (x', x_N) \in \mathbb{R}^{N-1} \times \mathbb{R}.$$

We now define a hypersurface in $\mathbb{R}^{N+1}$, which will provide a useful barrier:

DEFINITION 2.12. Given $Y \in \mathbb{R}^{N+1}$ with $|y_{N+1}| \leq \frac{1}{4}$ and $R$ large as above, let

(2.19) $\quad \mathbb{S}(Y, R) := \left\{ x \in \mathbb{R}^{N+1} \,\Big|\, x_{N+1} = g_{y_{N+1},R}\big(H_0(y_{N+1}) + |x - y| - R\big) \right\}.$

In the above definition, we will sometimes refer to $Y$ as the "center" and to $R$ as the "radius" of $\mathbb{S}$. For short, we also denote

(2.20) $\quad g_{\mathbb{S}(Y,R)}(x) := g_{y_{N+1},R}(H_0(y_{N+1}) + |x - y| - R),$

so that (2.19) becomes

$$\mathbb{S}(Y, R) = \left\{ x \in \mathbb{R}^{N+1} \,\Big|\, x_{N+1} = g_{\mathbb{S}(Y,R)}(x) \right\}.$$

Let us now prove that $g_{\mathbb{S}(Y,R)}$ is a strict supersolution in the viscosity sense (for the definition of viscosity super/sub/solutions, see, e.g., [**30**]):

PROPOSITION 2.13. Let $Y \in \mathbb{R}^{N+1}$ with $|y_{N+1}| \leq \frac{1}{4}$. Then, $g_{\mathbb{S}(Y,R)}$ is a strict supersolution of (1.5) in the viscosity sense at any $x \in \mathbb{R}^N$ for which $g_{\mathbb{S}(Y,R)}(x) \in [s_R, -1/2] \cap [1/2, 1)$.

Moreover, there are not smooth functions touching $g_{\mathbb{S}(Y,R)}$ by below at $x$ if $|g_{\mathbb{S}(Y,R)}(x)| = \frac{1}{2}$.

PROOF. We use the notation

$$s = g_{y_{N+1},R}(t) \quad \text{and} \quad t = H_0(y_{N+1}) + |x - y| - R$$

In this setting, we have to prove the desired supersolution property for $s_R \leq s \leq -1/2$ and for $1/2 \leq s < 1$.

Let us first consider the case $s = s_R$, that is $t \leq H_{y_{N+1},R}(s_R)$. In this case, $g_{y_{N+1},R}(t)$ is constantly equal to $s_R$, thus any paraboloid touching from above must have vanishing gradient at the contact point and negative definite Hessian matrix. Thus,

$$\Delta_p(g_{y_{N+1},R}(t)) = 0 < h_0'(s_R)$$

in the viscosity sense.

## 2. MODIFICATIONS OF THE POTENTIAL AND OF ONE-DIMENSIONAL SOLUTIONS

Let us now consider the case $H_{y_{N+1},R}(s_R) < t < H_{y_{N+1},R}(-\frac{1}{2})$ (that is, $s_R < s < -\frac{1}{2}$): in this case, $g_{y_{N+1},R}$ is smooth so that we can compute all the derivatives in the classic sense. As a matter of fact, by Lemma B.3,

$$g'_{y_{N+1},R}(t) = \left(\frac{p}{p-1} h_{y_{N+1},R}(g_{y_{N+1},R}(t))\right)^{1/p}$$

$$g''_{y_{N+1},R}(t) = \frac{\left(p h_{y_{N+1},R}(g_{y_{N+1},R}(t))\right)^{(2-p)/p}}{(p-1)^{2/p}} h'_{y_{N+1},R}(g_{y_{N+1},R}(t)).$$

Hence, exploiting Lemma B.2, we get

$$\Delta_p(g_{y_{N+1},R}(t)) = h'_{y_{N+1},R}(s) + \frac{N-1}{|x-y|}\left(\frac{p}{p-1} h_{y_{N+1},R}(s)\right)^{\frac{p-1}{p}}.$$

Notice that, from Lemma 2.10,

$$\begin{aligned} -C_1 \log R &\leq H_{y_{N+1},R}(s_R) < t = \\ &= H_0(y_{N+1}) + |x-y| - R \leq \\ &\leq \text{const} + |x-y| - R, \end{aligned}$$

and therefore, if $R$ is big enough,

$$(2.21) \qquad |x-y| \geq \frac{R}{2}.$$

Thus, we get

$$\begin{aligned} \Delta_p(g_{y_{N+1},R}(t)) &\leq h'_0(s) - \frac{\widehat{C}_0}{R} + \frac{2(N-1)}{R}\left(\frac{p}{p-1} h_0(s)\right)^{\frac{p-1}{p}} < \\ &< h'_0(s) \end{aligned}$$

provided that $\overline{C}_0$ (and so $\widehat{C}_0$) is chosen conveniently large. This proves the desired result for $H_{y_{N+1},R}(s_R) < t < H_{y_{N+1},R}(-\frac{1}{2})$.

If, on the other hand, $H_{y_{N+1},R}(\frac{1}{2}) < t < H_{y_{N+1},R}(1)$ (that is, $\frac{1}{2} < s < 1$), arguing in the same way, we get

$$\Delta_p(g_{y_{N+1},R}(t)) \leq h'_0(s) - \frac{\widehat{C}_0}{R} + \frac{2(N-1)}{R}\left(\frac{p}{p-1} h_{y_{N+1},R}(s)\right)^{\frac{p-1}{p}} < h'_0(s),$$

provided that $\widehat{C}_0$ is conveniently large. This completes the proof in the case $H_{y_{N+1},R}(\frac{1}{2}) < t < H_{y_{N+1},R}(1)$.

Up to now, we have therefore proved the desired result for

$$t \in \left(-\infty, H_{y_{N+1},R}(-\frac{1}{2})\right) \cup \left(H_{y_{N+1},R}(\frac{1}{2}), H_{y_{N+1},R}(1)\right),$$

that is, for

$$s \in \left[s_R, -\frac{1}{2}\right) \cup \left(\frac{1}{2}, 1\right).$$

To complete the proof of the claim, we have therefore to take now into account the case $|s| = 1/2$.

Let us now deal with the case $|s| = 1/2$. Recalling Lemma 2.6, by (2.2), we have

$$\lim_{s \to -\frac{1}{2}^-} H'_{y_{N+1},R}(s) < \lim_{s \to -\frac{1}{2}^+} H'_{y_{N+1},R}(s),$$

$$\lim_{s \to \frac{1}{2}^-} H'_{y_{N+1},R}(s) < \lim_{s \to \frac{1}{2}^+} H'_{y_{N+1},R}(s),$$

so that

(2.22)
$$\lim_{t \to H_{y_{N+1},R}(-\frac{1}{2})^-} g'_{y_{N+1},R}(t) > \lim_{t \to H_{y_{N+1},R}(-\frac{1}{2})^+} g'_{y_{N+1},R}(t),$$
$$\lim_{t \to H_{y_{N+1},R}(\frac{1}{2})^-} g'_{y_{N+1},R}(t) > \lim_{t \to H_{y_{N+1},R}(\frac{1}{2})^+} g'_{y_{N+1},R}(t).$$

Let now $w$ be a smooth function whose graph touches $\mathbb{S}(Y,R)$ by below at $(x_0, 1/2)$ (the case of touching at the $-1/2$-level is analogous). Let us consider the radial direction $\nu_0 = \frac{x_0 - y}{|x_0 - y|}$ and let us define, for $\theta \in \mathbb{R}$,

$$\mathfrak{g}(\theta) := g_{\mathbb{S}(Y,R)}(x_0 + \theta\nu_0) - w(x_0 + \theta\nu_0).$$

Then, by construction, $\mathfrak{g}(0) = 0 \leq \mathfrak{g}(\theta)$, therefore

$$0 \leq \lim_{\theta \to 0^+} \frac{\mathfrak{g}(\theta) - \mathfrak{g}(0)}{\theta} =$$
$$= \lim_{\theta \to 0^+} \frac{g_{y_{N+1},R}(t_0 + \theta) - g_{y_{N+1},R}(t_0)}{\theta} - \partial_{\nu_0} w(x_0) =$$
$$= \lim_{t \to H_{y_{N+1},R}(1/2)^+} g'_{y_{N+1},R}(t) - \partial_{\nu_0} w(x_0),$$

where

$$t_0 := H_0(y_{N+1}) + |x_0 - y| - R = H_{y_{N+1},R}(1/2).$$

By arguing in the same way, we also get that

$$0 \geq \lim_{\theta \to 0^-} \frac{\mathfrak{g}(\theta) - \mathfrak{g}(0)}{\theta} =$$
$$= \lim_{t \to H_{y_{N+1},R}(1/2)^-} g'_{y_{N+1},R}(t) - \partial_{\nu_0} w(x_0).$$

Thence,

$$\lim_{t \to H_{y_{N+1},R}(1/2)^-} g'_{y_{N+1},R}(t) \leq \partial_{\nu_0} w(x_0) \leq \lim_{t \to H_{y_{N+1},R}(1/2)^+} g'_{y_{N+1},R}(t),$$

which is a contradiction with (2.22). Therefore, no smooth function may touch $g_{\mathbb{S}(Y,R)}$ by below at $\pm 1/2$-level sets, showing, in particular, the claimed supersolution property. This completes the proof of Proposition 2.13. □

As a consequence of the above result, we show now that touching points between $\mathbb{S}(Y,R)$ and a subsolution of (1.5) may only occur when $|x_{N+1}| < 1/2$ (and this fact will be of great help in future computations, thanks to the explicit form of the barrier in $|x_{N+1}| < 1/2$):

COROLLARY 2.14. *Let $U \in C^1(\Omega)$ be a weak Sobolev subsolution of (1.5), with $|U| \leq 1$. Assume that $U \leq g_{\mathbb{S}(Y,R)}$ and that $U(x^\star) = g_{\mathbb{S}(Y,R)}(x^\star)$ for some $x^\star$ in the closure of $\Omega$. Then, either $x^\star \in \partial\Omega$ or $|g_{\mathbb{S}(Y,R)}(x^\star)| < 1/2$.*

## 2. MODIFICATIONS OF THE POTENTIAL AND OF ONE-DIMENSIONAL SOLUTIONS

PROOF. Let us assume that $x^\star \notin \partial\Omega$ We first prove that $g_{\mathbb{S}(Y,R)}(x^\star) \neq 1$. We argue by contradiction, assuming $g_{\mathbb{S}(Y,R)}(x^\star) = 1$. Notice that, due to Definition 2.5,

$$\lim_{s \to 1^-} \frac{dH_{y_{N+1},R}}{ds}(s) = \lim_{s \to 1^-} \frac{1}{\left(\frac{p}{p-1} h_{y_{N+1},R}(s)\right)^{1/p}} =$$

$$= \left(\frac{(p-1)R}{p}\right)^{1/p},$$

thus $g'_{y_{N+1},R}(t) > 0$ if $t = H_{y_{N+1},R}(1)$. Then, if

$$\nu^\star := \frac{x^\star - y}{|x^\star - y|}$$

is the radial direction, this yields

(2.23) $$\frac{\partial g_{\mathbb{S}(Y,R)}}{\partial \nu^\star}(x^\star) > 0.$$

However, since $U \in C^1(\mathbb{R}^N)$, $U(x^\star) = 1$ and $U \leq 1$,

(2.24) $$\nabla U(x^\star) = 0.$$

Similarly, since $U \leq g_{\mathbb{S}(Y,R)}$ and $U(x^\star) = g_{\mathbb{S}(Y,R)}(x^\star)$,

(2.25) $$\frac{\partial g_{\mathbb{S}(Y,R)}}{\partial \nu^\star}(x^\star) \leq \frac{\partial U}{\partial \nu^\star}(x^\star).$$

Thus, a contradiction easily follows from (2.23), (2.24) and (2.25), showing that $g_{\mathbb{S}(Y,R)}(x^\star) \neq 1$.

Also we claim that $g_{\mathbb{S}(Y,R)}(x^\star) \notin [s_R, -1/2] \cup [1/2, 1)$.

Let us first show that $g_{\mathbb{S}(Y,R)}(x^\star) \neq s_R$. We argue by contradiction and assume that $g_{\mathbb{S}(Y,R)}(x^\star) = s_R$. We recall that, by Definitions 2.11 and 2.12, $g_{\mathbb{S}(Y,R)}$ is constantly equal to $s_R$ in $B_r(y)$, with

$$r = r(R) := R + H_{y_{N+1},R}(s_R) - H_0(y_{N+1})$$

and[3] that $r > 0$ by (2.15). Then, there would be $\rho > 0$ so that

$$\Omega_\star := B_r(y) \cap B_\rho(x^\star) \subseteq \left\{g_{\mathbb{S}(Y,R)} = s_R\right\}.$$

Possibly taking $\rho$ smaller, we may assume also that

(2.26) $$\Omega_\star \subseteq \left\{U < -1 + \theta^*\right\}$$

and, since $x^\star \notin \partial\Omega$ by our assumption, that

(2.27) $$\Omega_\star \text{ is contained in the interior of } \Omega.$$

---

[3] Proceeding as done here and exploiting (2.16), one may also prove that $g_{\mathbb{S}(Y,R)}$ reaches the value 1 well inside the ball of radius $R + R^{\frac{1}{3}}/2$ around $y$.

Note that $U$ cannot coincide with $g_{\mathbb{S}(Y,R)}$ in $\Omega_\star$, otherwise

$$\begin{aligned}
0 &= -\int |\nabla U|^{p-2} \nabla U \cdot \nabla \varphi \\
&\geq \int h_0'(U) \varphi \\
&= \int h_0'(s_R) \varphi \\
&> 0,
\end{aligned}$$

for any non-negative smooth test-function $\varphi$ supported in $\Omega_\star$. Then, there exists $\bar{x} \in \Omega_\star$ and $\rho' > 0$ so that $U < g_{\mathbb{S}(Y,R)}$ in the interior of $B_{\rho'}(\bar{x}) \subseteq \Omega_\star$, but there exists $\check{x} \in \partial B_{\rho'}$ for which $U(\check{x}) = g_{\mathbb{S}(Y,R)}(\check{x})$. Setting

$$U^\star = s_R - U$$

it follows

$$U^\star = g_{\mathbb{S}(Y,R)} - U > 0$$

in $B_{\rho'}(\bar{x})$ and $U^\star(\check{x}) = 0$. Moreover,

(2.28) $$-\Delta_p U^\star = \Delta_p U \geq h_0'(U).$$

Hence, $h_0'(U) > 0$ in the light of (1.2) and (2.26). Thence, from (2.28), $-\Delta_p U^\star > 0$. Therefore, by Theorem B.6 (applied with $c = g = 0$),

(2.29) $$\partial_\nu U(\check{x}) = -\partial_\nu U^\star(\check{x}) > 0,$$

where

$$\nu := \frac{\check{x} - \bar{x}}{|\check{x} - \bar{x}|}$$

is the outer normal of $B_{\rho'}(\bar{x})$ at $\check{x}$.

On the contrary, note that $\check{x}$ is in the interior of the domain of $U$ thanks to (2.27), and so, since $U$ touches $g_{\mathbb{S}(Y,R)}$ at $\check{x}$ and $\nabla g_{\mathbb{S}(Y,R)}$ vanishes on $\{g_{\mathbb{S}(Y,R)} = s_R\}$, we have that

$$\partial_\nu U(\check{x}) = \partial_\nu g_{\mathbb{S}(Y,R)}(\check{x}) = 0,$$

against (2.29). This contradiction shows that $x^\star$ does not lie in $\{g_{\mathbb{S}(Y,R)} = s_R\}$.

Let us now prove that

(2.30) $$g_{\mathbb{S}(Y,R)}(x^\star) \notin \left(s_R, -\frac{1}{2}\right).$$

First, we recall that $g_{\mathbb{S}(Y,R)}$ is smooth with non-vanishing gradient in

$$\Omega_0 := \left\{ g_{\mathbb{S}(Y,R)}(x) \in \left(s_R, -\frac{1}{2}\right) \right\},$$

thence it is a classical strict supersolution of (1.5) in $\Omega_0$. This implies that $U$ cannot coincide with $g_{\mathbb{S}(Y,R)}$ in $\{g_{\mathbb{S}(Y,R)}(x^\star) \in (s_R, -\frac{1}{2})\}$, otherwise,

$$\begin{aligned}\int h_0'(g_{\mathbb{S}(Y,R)})\varphi &> -\int |\nabla g_{\mathbb{S}(Y,R)}|^{p-2}\nabla g_{\mathbb{S}(Y,R)} \cdot \nabla\varphi = \\ &= -\int |\nabla U|^{p-2}\nabla U \cdot \nabla\varphi \geq \\ &\geq \int h_0'(U)\varphi \\ &= \int h_0'(g_{\mathbb{S}(Y,R)})\varphi,\end{aligned}$$

for any non-negative smooth test-function $\varphi$ supported in $\Omega_0$, which is an obvious contradiction. Therefore, since $U$ and $g_{\mathbb{S}(Y,R)}$ do not agree in $\Omega_0$ and $\nabla g_{\mathbb{S}(Y,R)}$ never vanishes there, we can exploit Corollary B.5 and get that $U < g_{\mathbb{S}(Y,R)}$ in $\Omega_0$. Thence, no touching point may occur in $\{g_{\mathbb{S}(Y,R)}(x) \in (s_R, -\frac{1}{2})\}$, proving (2.30).

Also, $g_{\mathbb{S}(Y,R)}(x^\star) \neq -\frac{1}{2}$ by Proposition 2.13. The fact that $g_{\mathbb{S}(Y,R)}(x^\star) \notin [1/2, 1)$ follows with similar arguments. □

We will now define another hypersurface in $\mathbb{R}^{N+1}$, which will be denoted by $\widetilde{\mathbb{S}}(Y,R)$ and we investigate its relation with $\mathbb{S}(Y,R)$. While $\mathbb{S}(Y,R)$ is continuous but not smooth, $\widetilde{\mathbb{S}}(Y,R)$ will be smooth, and thus it will be easier to deal with during the calculations. Also, the two surfaces will coincide in $\{|x_{N+1}| \leq \frac{1}{2}\}$ and $\mathbb{S}(Y,R)$ will always stay below $\widetilde{\mathbb{S}}(Y,R)$. Therefore, $\widetilde{\mathbb{S}}(Y,R)$ will provide, in some sense, a sharp barrier for $\mathbb{S}(Y,R)$ which will be more explicit to treat. Let us now approach the definition of the hypersurface $\widetilde{\mathbb{S}}(Y,R)$.

Setting

$$(2.31) \qquad \widetilde{H}_{s_0,R}(s) := H_0(s) - \frac{\overline{C}_0}{2R}(s-s_0)^2$$

then, by (2.12), we have

$$(2.32) \qquad \frac{d}{ds}\left(\widetilde{H}_{s_0,R}(s)\right) = \frac{1}{(\frac{p}{p-1}h_0(s))^{\frac{1}{p}}} - \frac{\overline{C}_0}{R}(s-s_0) > 0$$

for $|s| \leq \frac{3}{4}$, provided that $R$ is big enough. Therefore, for $|s| \leq \frac{3}{4}$ we have that $\widetilde{H}_{s_0,R}$ is strictly increasing and we can give the following

DEFINITION 2.15. We define

$$\rho_{s_0,R}(t) : \left[\widetilde{H}_{s_0,R}(-\frac{3}{4}), \widetilde{H}_{s_0,R}(\frac{3}{4})\right] \longrightarrow \left[-\frac{3}{4}, \frac{3}{4}\right]$$

by

$$(2.33) \qquad \rho_{s_0,R}(t) := \widetilde{H}_{s_0,R}^{-1}(t).$$

Moreover, given $Y \in \mathbb{R}^{N+1}$ with $|y_{N+1}| \leq \frac{1}{4}$ and $R$ large as above, we define

$$(2.34) \qquad \widetilde{\mathbb{S}}(Y,R) := \{x \in \mathbb{R}^{N+1} \,|\, x_{N+1} = \rho_{y_{N+1},R}(H_0(y_{N+1}) + |x-y| - R)\}$$

As done on page 14, it is convenient to introduce the notation

(2.35) $$g_{\widetilde{\mathbb{S}}(Y,R)}(x) := \rho_{y_{N+1},R}(H_0(y_{N+1}) + |x-y| - R),$$

so that (2.34) becomes

$$\widetilde{\mathbb{S}}(Y,R) = \left\{ x \in \mathbb{R}^{N+1} \,\Big|\, x_{N+1} = g_{\widetilde{\mathbb{S}}(Y,R)}(x) \right\}.$$

Notice also that, by construction, for any $x$ for which $g_{\widetilde{\mathbb{S}}(Y,R)}$ is defined[4], we have that

(2.37) $$|g_{\widetilde{\mathbb{S}}(Y,R)}(x)| \leq \frac{3}{4}.$$

Moreover, a straightforward computation gives that

(2.38) $$\widetilde{H}_{s_0,R_1}(s) - \widetilde{H}_{s_0,R_2}(s) = \frac{\overline{C_0}}{2}(s-s_0)^2 \left( \frac{1}{R_2} - \frac{1}{R_1} \right),$$

for any $|s| \leq 3/4$. Also, if $x$ is in the domain of $g_{\widetilde{\mathbb{S}}(Y,R)}$, then $x$ and $y$ must be suitably far from each other, as next result points out:

LEMMA 2.16. *Let $x \in \mathbb{R}^N$ be so that $g_{\widetilde{\mathbb{S}}(Y,R)}$ is well defined (that is, let $x$ be such that (2.36) holds). Then,*

$$|x-y| \geq R - C,$$

*for a suitable universal constant $C > 0$.*

PROOF. From (2.37), we get that

$$s := g_{\widetilde{\mathbb{S}}(Y,R)}(x) \in [-3/4,\, 3/4].$$

So, by means of (2.31) and (2.33),

$$|x-y| = R - H_0(y_{N+1}) + H_0(s) - \frac{\overline{C_0}}{2R}(s - y_{N+1})^2 \geq$$
$$\geq R - \text{const}.$$

$\square$

Let us now show that the surface $\mathbb{S}(Y,R)$ coincides with $\widetilde{\mathbb{S}}(Y,R)$ in the set $|x_{N+1}| \leq \frac{1}{2}$ and that $\mathbb{S}(Y,R)$ stays below $\widetilde{\mathbb{S}}(Y,R)$ at all other points where $\widetilde{\mathbb{S}}(Y,R)$ is defined:

LEMMA 2.17. *If $|s| \leq 1/2$, then*

(2.39) $$H_{s_0,R}(s) = \widetilde{H}_{s_0,R}(s).$$

*If $\frac{1}{2} < |s| < \frac{3}{4}$, then*

(2.40) $$H_{s_0,R}(s) > \widetilde{H}_{s_0,R}(s).$$

*Also, let $x \in \mathbb{R}^N$ be so that $g_{\widetilde{\mathbb{S}}(Y,R)}$ is well defined (that is, let $x$ be such that (2.36) holds). Then,*

(2.41) $$g_{\widetilde{\mathbb{S}}(Y,R)}(x) \geq g_{\mathbb{S}(Y,R)}(x).$$

---

[4] I.e., for any $x$ so that

(2.36) $$H_0(y_{N+1}) + |x-y| - R \in \left[ \widetilde{H}_{s_0,R}(-\frac{3}{4}),\, \widetilde{H}_{s_0,R}(\frac{3}{4}) \right].$$

## 2. MODIFICATIONS OF THE POTENTIAL AND OF ONE-DIMENSIONAL SOLUTIONS

*Furthermore, if* $|g_{\widetilde{\mathbb{S}}(Y,R)}(x)| \leq \frac{1}{2}$, *then*

$$\tag{2.42} g_{\widetilde{\mathbb{S}}(Y,R)}(x) = g_{\mathbb{S}(Y,R)}(x).$$

PROOF. We use the notation $s := g_{\widetilde{\mathbb{S}}(Y,R)}$. Let $|s| \leq 1/2$. By Definitions 2.5 and 2.8, recalling also (2.2), we have that

$$\tag{2.43} \begin{aligned} H_{s_0,R}(s) &= H_0(s_0) + \int_{s_0}^{s} \frac{d\zeta}{\left(\frac{p}{p-1}h_0(s)\right)^{\frac{1}{p}}} - \int_{s_0}^{s} \frac{\overline{C_0}}{R}(s - s_0) = \\ &= H_0(s) - \frac{\overline{C_0}}{2R}(s - s_0)^2 = \\ &= \widetilde{H}_{s_0,R}(s). \end{aligned}$$

This and (2.31) prove (2.39) and (2.42). We now prove (2.40) and (2.41). Let us consider only the case $s \in [-3/4, -1/2]$, the case $s \in [1/2, 3/4]$ being analogous. In this case, $s < s_0$, and thus, exploiting Lemma 2.6, we get

$$\begin{aligned} H_{s_0,R}(s) &= H_0(s_0) - \int_s^{s_0} \frac{1}{\left(\frac{p}{p-1}h_{s_0,R}\right)^{1/p}} > \\ &> H_0(s_0) - \int_s^{s_0} \frac{1}{\left(\frac{p}{p-1}\varphi\right)^{1/p}} = \\ &= H_0(s_0) - \int_s^{s_0} \frac{1}{\left(\frac{p}{p-1}h_0\right)^{1/p}} - \frac{\overline{C_0}}{2R}(s - s_0)^2 = \\ &= H_0(s) - \frac{\overline{C_0}}{2R}(s - s_0)^2, \end{aligned}$$

proving (2.40) and (2.41). $\square$

Since, by construction, the function $g_{\widetilde{\mathbb{S}}(Y,R)}$ defined above is smooth (and, due to (2.32, its gradient never vanishes), we can compute its derivatives (and its $p$-Laplacian) in the classic sense. In particular, we can sharply estimate how far $g_{\widetilde{\mathbb{S}}(Y,R)}$ is from being a solution of (1.5), thanks to the following result:

PROPOSITION 2.18. *Let* $Y \in \mathbb{R}^{N+1}$ *with* $|y_{N+1}| \leq \frac{1}{4}$. *Then, there exists a positive universal constant* $C > 0$ *such that*

$$h_0'\left(g_{\widetilde{\mathbb{S}}(Y,R)}(x)\right) - \frac{C}{R} \leq \Delta_p g_{\widetilde{\mathbb{S}}(Y,R)}(x) \leq h_0'\left(g_{\widetilde{\mathbb{S}}(Y,R)}(x)\right) + \frac{C}{R},$$

*for any $x$ for which $g_{\widetilde{\mathbb{S}}(Y,R)}$ is defined (i.e., for any $x$ so that (2.36) holds).*

PROOF. We will use the notation

$$\tag{2.44} \begin{aligned} t &:= H_0(y_{N+1}) + |x - y| - R \quad \text{and} \\ s &:= \rho_{y_{N+1},R}(t) = g_{\widetilde{\mathbb{S}}(Y,R)}(x). \end{aligned}$$

Let us note that, from (2.31) and (2.2),

$$\tag{2.45} \frac{d}{dt}\rho_{y_{N+1},R}(t) = \frac{1}{H_0'(s) - \frac{\overline{C_0}}{R}(s - y_{N+1})} = \left(\frac{p}{p-1}\varphi(s)\right)^{\frac{1}{p}}.$$

Hence, differentiating again, a straightforward calculation leads to

$$\frac{d^2}{dt^2}\rho_{y_{N+1},R}(t) = \frac{(p\,\varphi(s))^{\frac{2-p}{p}}}{(p-1)^{\frac{2}{p}}}\,\varphi'(s). \tag{2.46}$$

Therefore, by Lemma B.2,

$$\Delta_p(\rho_{y_{N+1},R}(t)) = \varphi'(s) + \frac{N-1}{|x-y|}\left(\frac{p}{p-1}\,\varphi(s)\right)^{\frac{p-1}{p}}. \tag{2.47}$$

Furthermore, note that, by differentiating (2.2), one obtains

$$-\frac{1}{p}\left(\frac{\frac{p}{p-1}\varphi'(s)}{\left(\frac{p}{p-1}\varphi(s)\right)^{\frac{p+1}{p}}}\right) = -\frac{1}{p}\left(\frac{\frac{p}{p-1}h_0'(s)}{\left(\frac{p}{p-1}h_0(s)\right)^{\frac{p+1}{p}}}\right) - \frac{\overline{C}_0}{R} \tag{2.48}$$

so that

$$\varphi'(s) = \left(\frac{\varphi(s)}{h_0(s)}\right)^{\frac{p+1}{p}} h_0'(s) + \frac{p\overline{C}_0}{R}\left(\frac{p}{p-1}\right)^{\frac{1}{p}}(\varphi(s))^{\frac{p+1}{p}}, \tag{2.49}$$

whence, from (2.1),

$$\varphi'(s) = h_0'(s) +$$
$$+ \frac{1}{R}\left\{h_0'(s)R\left[\left(\frac{R}{R-a}\right)^{p+1} - 1\right] + p\overline{C}_0\left(\frac{p}{p-1}\right)^{\frac{1}{p}}(\varphi(s))^{\frac{p+1}{p}}\right\}, \tag{2.50}$$

with

$$a = a(s) := \overline{C}_0\,(s-s_0)\left(\frac{p}{p-1}\,h_0(s)\right)^{1/p}.$$

Using now the fact that

$$\lim_{x\to 0^+}\left|\frac{\left(\frac{1}{1-xa}\right)^{p+1} - 1}{x}\right| = |(p+1)a| < +\infty,$$

we obtain, if $R$ is suitably large, that

$$\left|h_0'(s)R\left[\left(\frac{R}{R-a}\right)^{p+1} - 1\right] + p\overline{C}_0\left(\frac{p}{p-1}\right)^{\frac{1}{p}}(\varphi(s))^{\frac{p+1}{p}}\right| \leq \text{const}, \tag{2.51}$$

so that, from (2.47) and the fact that $\varphi \geq 0$,

$$h_0'(s) - \frac{\text{const}}{R} \leq \Delta_p(\rho_{y_{N+1},R}(t)),$$

which proves one side of the claimed inequality.

The other side of the inequality is obtained by arguing in the same way, making use also of Lemma 2.16. □

For further reference, we point out some easy calculations on the above barrier:

LEMMA 2.19. *At any $x$ for which $g_{\widetilde{\mathbb{S}}(Y,R)}$ is defined, we have that*

$$\left(H_0'\left(g_{\widetilde{\mathbb{S}}(Y,R)}(x)\right) - \frac{\overline{C}_0}{R}\left(g_{\widetilde{\mathbb{S}}(Y,R)}(x) - y_{N+1}\right)\right)\nabla g_{\widetilde{\mathbb{S}}(Y,R)}(x) =$$
$$= \frac{x-y}{|x-y|}. \tag{2.52}$$

## 2. MODIFICATIONS OF THE POTENTIAL AND OF ONE-DIMENSIONAL SOLUTIONS 23

Also, there exists a universal constant $C > 0$ so that

$$\left| H_0' \left( g_{\widetilde{\mathbb{S}}(Y,R)}(x) \right) \left| \nabla g_{\widetilde{\mathbb{S}}(Y,R)}(x) \right| - 1 \right| \leq \frac{C}{R} \tag{2.53}$$

and

$$\left| H_0'' \left( g_{\widetilde{\mathbb{S}}(Y,R)}(x) \right) \partial_i g_{\widetilde{\mathbb{S}}(Y,R)}(x) \partial_j g_{\widetilde{\mathbb{S}}(Y,R)}(x) + \right.$$
$$\left. + H_0' \left( g_{\widetilde{\mathbb{S}}(Y,R)}(x) \right) \partial_{ij} g_{\widetilde{\mathbb{S}}(Y,R)}(x) \right| \leq$$
$$\leq \frac{C}{R}, \tag{2.54}$$

provided that $R$ is large enough.

PROOF. Using the notation in (2.44) and the first equality in (2.45), the claim in (2.52) easily follows. From (2.52) and (2.37), one easily gets (2.53).

Let us now prove (2.54). From (2.31) and (2.33), we have that

$$\tau = H_0\big(\rho_{y_{N+1},R}(\tau)\big) - \frac{\overline{C}_0}{2R} \big(\rho_{y_{N+1},R}(\tau) - y_{N+1}\big)^2,$$

for any $\tau$ for which $\rho_{y_{N+1},R}$ is defined; then, differentiating twice and recalling (2.37),

$$\left| H_0''(\rho_{y_{N+1},R}(\tau)) \left(\rho'_{y_{N+1},R}(\tau)\right)^2 + \right.$$
$$\left. + H_0'(\rho_{y_{N+1},R}(\tau)) \rho''_{y_{N+1},R}(\tau) \right| \leq \frac{\text{const}}{R}. \tag{2.55}$$

Furthermore, differentiating twice the relation

$$g_{\widetilde{\mathbb{S}}(Y,R)}(x) = \rho_{y_{N+1},R}\Big(H_0(y_{N+1}) + |x - y| - R\Big),$$

a direct computation gives

$$\partial_{ij} g_{\widetilde{\mathbb{S}}(Y,R)}(x) =$$
$$= R_{ij} \rho''_{y_{N+1},R}(t) + \left(\frac{\delta_{ij}}{|x - y|} - \frac{R_{ij}}{|x - y|}\right) \rho'_{y_{N+1},R}(t),$$

where we define, for short,

$$R_{ij} := \frac{(x_i - y_i)(x_j - y_j)}{|x - y|^2}.$$

In other words,

$$\partial_{ij} g_{\widetilde{\mathbb{S}}(Y,R)}(x) = R_{ij} \rho''_{y_{N+1},R}(t) + S_{ij} \tag{2.56}$$

for a suitable $S_{ij}$ satisfying $|S_{ij}| \leq \text{const}/|x - y|$. Therefore, using (2.56) and the fact that

$$\partial_i g_{\widetilde{\mathbb{S}}(Y,R)}(x) = \rho'_{y_{N+1},R}(t) \frac{x_i - y_i}{|x - y|}, \tag{2.57}$$

we get that

$$H_0''\left(g_{\widetilde{\mathbb{S}}(Y,R)}(x)\right) \partial_i g_{\widetilde{\mathbb{S}}(Y,R)}(x) \partial_j g_{\widetilde{\mathbb{S}}(Y,R)}(x) +$$
$$+ H_0'\left(g_{\widetilde{\mathbb{S}}(Y,R)}(x)\right) \partial_{ij} g_{\widetilde{\mathbb{S}}(Y,R)}(x) =$$
$$= H_0''(\rho_{y_{N+1},R}(t)) \left(\rho_{y_{N+1},R}(t)\right)^2 R_{ij} +$$
$$+ H_0'(\rho_{y_{N+1},R}(t)) \left(R_{ij}\rho_{y_{N+1},R}''(t) + S_{ij}\right) =$$
$$= R_{ij}\left[H_0''\left(\rho_{y_{N+1},R}(t)\right)\left(\rho_{y_{N+1},R}(t)\right)^2 +\right.$$
$$\left.+ H_0'\left(\rho_{y_{N+1},R}(t)\right)\rho_{y_{N+1},R}''(t)\right] + T_{ij},$$

with $|T_{ij}| \leq \text{const}/|x-y|$. Thus, from (2.55),

$$\left|H_0''\left(g_{\widetilde{\mathbb{S}}(Y,R)}(x)\right)\partial_i g_{\widetilde{\mathbb{S}}(Y,R)}(x)\partial_j g_{\widetilde{\mathbb{S}}(Y,R)}(x) +\right.$$
$$\left.+ H_0'\left(g_{\widetilde{\mathbb{S}}(Y,R)}(x)\right)\partial_{ij} g_{\widetilde{\mathbb{S}}(Y,R)}(x)\right| \leq$$
$$\leq \text{const}\left(\frac{1}{R} + \frac{1}{|x-y|}\right).$$

Therefore, (2.54) is proved thanks to Lemma 2.16. □

We now recall a result, proved in [30], concerning another barrier which will be used in the course of the proof of the main results:

LEMMA 2.20. *There exist universal constants $\bar{l} > 1$ and $0 < \bar{c} \leq 1/2$, so that, if $l \geq \bar{l}$, we can find $T_l \in [\bar{c}l, l/2]$ and a nondecreasing function*

$$g_l \in C^0(-\infty, T_l) \cap C^{1,1}(-\infty, 0) \cap C^2((-\bar{c}l, T_l) \setminus \{0\})$$

*which is constant in $(-\infty, -l/2]$, with $g_l' > 0$ in $[-\bar{c}l, T_l]$, satisfying $g_l(0) = 0$, $g_l(T_l) = 1$ and such that, if we define*

(2.58) $$\Psi^{y,l}(x) := g_l(|x-y| - l),$$

*then $\Psi^{y,l}(x)$ is a strict supersolution of (1.5) in the viscosity sense, in $B_{T_l+l}(y) \setminus \partial B_l(y)$.*

*More precisely, $g_l$ is constructed as follows. There exists a suitable constants $0 < c_1 < \overline{C}_1, \overline{C}_2$ so that, if we define*

$$s_l := e^{-\bar{c}_1 l},$$

$$h_l(s) := \begin{cases} h_0(s) - h_0(s_l - 1) - \frac{\overline{C}_2}{l}\left((1+s)^p - s_l^p\right) \\ \quad \text{if} \quad (s_l - 1) < s < 0 \\ h_0(s) + h_0(1 - s_l) + \frac{\overline{C}_2}{l}\left((1-s)^p + s_l^{(p-1)}(1-s)\right) \\ \quad \text{if} \quad 0 \leq s < 1, \end{cases}$$

$$H_l(s) := \int_0^s \frac{(p-1)^{\frac{1}{p}}}{(p\,h_l(\zeta))^{\frac{1}{p}}}\, d\zeta,$$

$$H_0(s) := \int_0^s \frac{(p-1)^{\frac{1}{p}}}{(p\,h_0(\zeta))^{\frac{1}{p}}}\, d\zeta, \quad \textit{for any } s \in (-1,1),$$

*then the following holds:*

(i) $h_l(s) > 0$ in $(s_l - 1) < s < 1$; in particular, $H_l$ is well defined and strictly increasing for $(s_l - 1) < s < 1$ and thus we may define $g_l(t) := H_l^{-1}(t)$ for $t \in H_l(s_l - 1, 1)$;

(ii) $g_l(t)$ is defined to be constantly equal to $s_l - 1$ for $t \leq H_l(s_l - 1)$;

(iii) the following estimates on $H_l$ hold:

$$H_l(1) \leq \frac{l}{2}; \tag{2.59}$$

$$H_l(s_l - 1) \geq -\frac{l}{2}; \tag{2.60}$$

$$H_0(s) \leq H_l(s) - \frac{\overline{C}_1}{l} \log(1 - |s|) \quad \forall |s| < 1 - e^{-\frac{\overline{c}_1 l}{2}}; \tag{2.61}$$

$$H_l(1 - e^{-\frac{\overline{c}_1 l}{2}}) \geq \overline{c}\, l; \tag{2.62}$$

$$H_l(e^{-\frac{\overline{c}_1 l}{2}} - 1) \leq -\overline{c}\, l. \tag{2.63}$$

A detailed proof of Lemma 2.20 is contained in [**30**] (see, in particular, Lemma 5.1 there).

We now point out some properties of the touching points between the barrier $\Psi^{y,l}$ and a (sub)solution of (1.5). To this end, we notice that, by construction, the radially increasing function $\Psi^{y,l}$ built in Lemma 2.20 is so that:

- $\Psi^{y,l}$ is defined in $B_{T_l + l}(y)$, and it is greater than $s_l - 1$;
- there exists $\rho_l \in [\overline{c}l, l/2]$ so that the only critical points of $\Psi^{y,l}$ are in $B_{l - \rho_l}(y)$, where $\Psi^{y,l}$ is flat;
- $\Psi^{y,l} = 0$ on $\partial B_l(y)$.

The geometry of such spheres is related with possible touching points, as next result shows:

LEMMA 2.21. *Fix $y \in \mathbb{R}^N$ and let $l > 0$ be suitably large. Let $u$ be a weak Sobolev subsolution of (1.5) in some domain $\Omega$. Suppose that $u \in C^1(\Omega)$ and that $|u| \leq 1$. Then the following results hold:*

- *If $\Psi^{y,l}$ touches the graph of $u$ from above at some point $x^\star$ in the closure of $\Omega \cap B_{T_l + l}(y)$, then, either $x^\star \in \partial\Omega$ or $u(x^\star) = \Psi^{y,l}(x^\star) = 0$.*
- *If*

$$B_{l + T_l}(y) \subset \{x \in \Omega \mid u(x) \leq -1 + \theta^*\},$$

*then,*

$$u(x) < \Psi^{y,l}(x),$$

*for any $x \in B_{l + T_l}(y)$.*

For the proof of Lemma 2.21, we refer to [**30**] (see, in particular, Lemma 6.2 and Corollary 6.4 there). We now notice that a statement analogous to Lemma 2.20 holds for a subsolution (instead of supersolution) property:

LEMMA 2.22. *There exist universal constants $\overline{l} > 1$ and $0 < \overline{c} \leq 1/2$, so that, if $l \geq \overline{l}$, we can find $T_l \in [\overline{c}l, l/2]$ and a nondecreasing function*

$$\widetilde{g}_l \in C^0(-T_l, +\infty) \cap C^{1,1}(0, +\infty) \cap C^2((-T_l, \overline{c}l) \setminus \{0\})$$

*which is constant in $[l/2, +\infty)$, with $\widetilde{g}_l' > 0$ in $[-T_l, \overline{c}l]$, satisfying $\widetilde{g}_l(0) = 0$, $\widetilde{g}_l(-T_l) = -1$ and such that, if we define*

$$\widetilde{\Psi}^{y,l}(x) := \widetilde{g}_l(l - |x - y|),$$

then $\widetilde{\Psi}^{y,l}(x)$ is a strict subsolution of (1.5) in the viscosity sense, in $B_{T_l+l}(y) \setminus \partial B_l(y)$.

Also, if we define $\widetilde{h}_0(s) := h_0(-s)$ and

$$\widetilde{H}_0(t) := \int_0^t \frac{(p-1)^{1/p}\, d\zeta}{(p\widetilde{h}_0(\zeta))^{1/p}},$$

then, there exists a strictly increasing function $\widetilde{H}_l$ and a positive function $h_l$, such that

$$\widetilde{H}_l'(s) = \frac{(p-1)^{1/p}}{(p\widetilde{h}_l(s))^{1/p}},$$

so that the following holds:

(i) $\widetilde{h}_l(s)$ is defined and strictly positive in $-1 < s < 1 - s_l$; $\widetilde{H}_l$ is defined and strictly increasing for $-1 < s < 1 - s_l$; $\widetilde{g}_l(t) = \widetilde{H}_l^{-1}(t)$ for $t \in \widetilde{H}_l(1, 1-s_l)$;

(ii) $\widetilde{g}_l(t)$ is constantly equal to $1 - s_l$ for $t \geq \widetilde{H}_l(1 - s_l)$;

(iii) the following estimates on $\widetilde{H}_l$ hold:

$$\widetilde{H}_l(-1) \geq -\frac{l}{2},$$

$$\widetilde{H}_l(1 - s_l) \leq \frac{l}{2},$$

$$\widetilde{H}_0(s) \geq \widetilde{H}_l(s) + \frac{\overline{C}_1}{l}\log(1 - |s|) \quad \forall |s| < 1 - e^{-\frac{\overline{c}_1 l}{2}},$$

$$\widetilde{H}_l(-1 + e^{-\frac{\overline{c}_1 l}{2}}) \leq -\bar{c}\, l,$$

$$H_l(1 - e^{-\frac{\overline{c}_1 l}{2}}) \geq \bar{c}\, l.$$

PROOF. Notice that $\widetilde{h}_0$ satisfies the same assumption as $h_0$, thus, we can use Lemma 2.20 with $\widetilde{h}_0$ replacing $h_0$: let us denote by $h_l^\sharp$, $H_l^\sharp$ and $g_l^\sharp$ the functions thus obtained via Lemma 2.20 with $\widetilde{h}_0$ replacing $h_0$. Then, define

$$\widetilde{h}_l(s) := h_l^\sharp(-s).$$

Thus,

$$\widetilde{H}_l(s) := \int_0^s \frac{(p-1)^{1/p}}{(p\widetilde{h}_l)^{1/p}} = -H_l^\sharp(-s)$$

and therefore

$$\widetilde{g}_l(t) = -g_l^\sharp(-t).$$

With this, let us show that $\widetilde{\Psi}^{y,l}$ is a strict viscosity subsolution of (1.5) outside $\partial B_l(y)$. Indeed, if $\phi$ is a smooth function touching $\widetilde{\Psi}^{y,l}$ from above at $x^\star \notin \partial B_l(y)$, then $\varphi(x) := -\phi(x)$ touches from below at the point $x^\star$ the strict viscosity supersolution

$$-\widetilde{\Psi}^{y,l}(x) = g_l^\sharp(|x - y| - l).$$

Thus, by Lemma 2.20,

$$\Delta_p \phi = -\Delta_p \varphi > -(h_l^\sharp)'(\varphi) = \widetilde{h}_l'(-\varphi) = \widetilde{h}_l'(\phi)$$

at the point $x^\star$, which shows the desired subsolution property of $\widetilde{\Psi}^{y,l}$.

It is easy to see that $\widetilde{g}_l$ and $\widetilde{H}_l$ also enjoy the properties listed above. □

## 2. MODIFICATIONS OF THE POTENTIAL AND OF ONE-DIMENSIONAL SOLUTIONS

We are now in the position of showing that minimizers are trapped in between the functions constructed in Lemmata 2.20-2.22 (an exponential decay was also pointed out in §9 of [**30**]):

LEMMA 2.23. *Let $l > 0$, $\theta > 0$. Let $u$ be a local minimizer for $\mathcal{F}$ in*

$$\left\{ (x', x_N) \in \mathbb{R}^{N-1} \times \mathbb{R} \,\Big|\, |x'| < l,\ |x_N| < l \right\}.$$

*Assume that $|u| \leq 1$, that $\theta/l$ is suitably small, that $u(0) = 0$, that $u(x) > 0$ if $x_N \geq \theta$ and that $u(x) < 0$ if $x_N \leq -\theta$. Then, there exist suitable constants $\kappa, \widetilde{\kappa}, \hat{\kappa} \in (0, 1]$ so that*

$$\widetilde{g}_{\widetilde{\kappa}l}(x_N - \theta) \leq u(x) \leq g_{\kappa l}(x_N + \theta),$$

*at any point $x \in [-\hat{\kappa}l, \hat{\kappa}l]^N$, provided that the functions above are defined at $x$.*

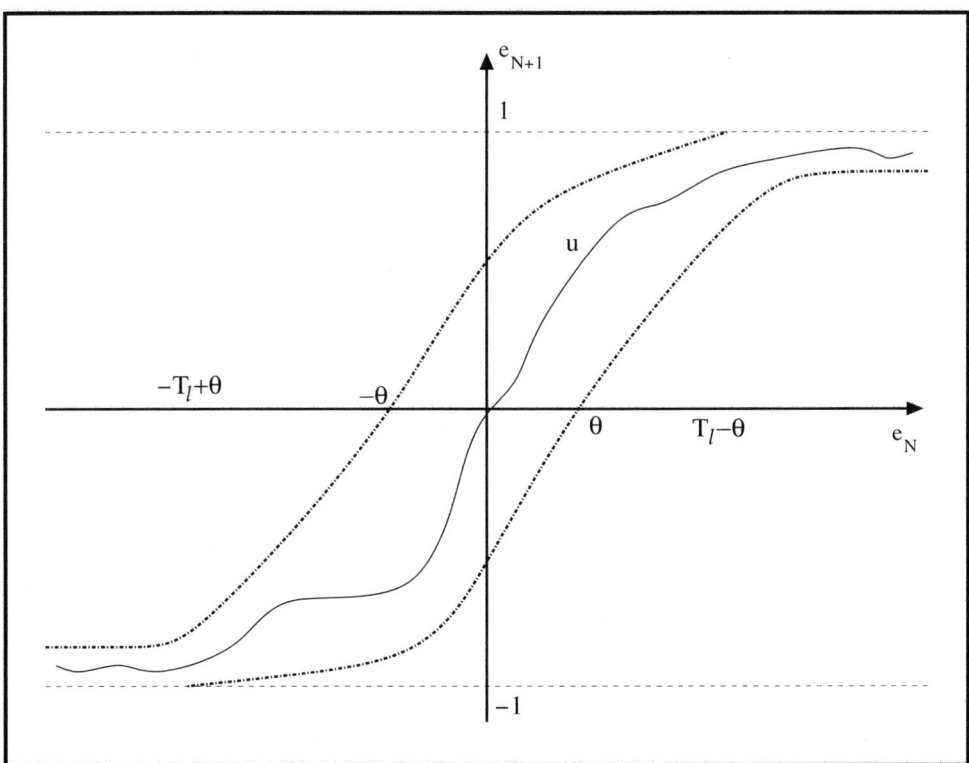

**Trapping a minimizer between two barriers, as in Lemma 2.23**

PROOF. We proof the latter inequality, the first one being analogous. From Theorem 1.1 of [**28**] and the fact that $\{u = 0\} \subseteq \{|x_N| \leq \theta\}$, it easily follows that

(2.64) $\qquad u(x) < -1 + \theta^\star$ for any $x \in \mathbb{R}^N$ with $x_N \leq \kappa^\star$ and $|x'| \leq l/2$,

for some constant $\kappa^\star$ (which may depend on $\theta^\star$ and other universal constants).

Then, from (2.64) and the second item in Lemma 2.21,

$$u(x) \leq \Psi^{y,\kappa l}(x),$$

where $y := (0, \ldots, 0, -l/2)$. Let now $e \in S^{N-1}$, with $e_N > 0$ and let us slide $\Psi^{y,\kappa l}$ in the $e$-direction until it touches $u$. Notice indeed that there exists a suitable constant $\bar{c} \in (0,1)$ so that if

(2.65) $$e_N \geq \bar{c},$$

then $\Psi^{y+te,\kappa l}$ does touch $u$ for some $t = t(e) > 0$ at some point $x^\star = x^\star(e) \in [-l,l]^N$, that is

(2.66) $$u(\mathfrak{x}) \leq \Psi^{y+te,\kappa l}(\mathfrak{x}) = g_{\kappa l}(|\mathfrak{x} - (y+te)| - \kappa l),$$

for any $\mathfrak{x} \in B_{T_{\kappa l}+\kappa l}(y+te)$, being the latter the domain where $\Psi^{y+te,\kappa l}$ is defined, and

$$u(x^\star) = \Psi^{y+te,\kappa l}(x^\star).$$

In the light of the first item in Lemma 2.21, we have that

$$u(x^\star) = \Psi^{y+te,\kappa l}(x^\star) = 0$$

and so, since, by our hypotheses $\{\mathfrak{x} \mid u(\mathfrak{x}) = 0\} \subseteq \{|\mathfrak{x}_N| \leq \theta\}$, we have that

$$|x_N^\star| \leq \theta.$$

Let us now consider, for $d \geq 0$, the point

$$\mathfrak{x} = \mathfrak{x}(e) := x^\star + d e_N.$$

Then,

$$\begin{aligned}
|\mathfrak{x} - (y+te)| &\leq |\mathfrak{x} - x^\star| + |x^\star - (y+te)| = \\
&= d + \kappa l = \\
&= \mathfrak{x}_N - x_N^\star + \kappa l \\
&\leq \mathfrak{x}_N + \theta + \kappa l.
\end{aligned}$$

Therefore, since $g_l$ is increasing, (2.66) implies that

(2.67) $$u(\mathfrak{x}) \leq g_{\kappa l}(\mathfrak{x}_N + \theta),$$

provided $\mathfrak{x} \in B_{T_{\kappa l}+\kappa l}(y+te)$.

With this inequality, we are now in the position of choosing $e$ here above in order to infer the desired result. We proceed in the following way: take $x = (x', x_N) \in [-\hat{\kappa}l, \hat{\kappa}l]^{N-1} \times [-\hat{\kappa}l, \hat{\kappa}l]$ and consider two cases.

If

(2.68) $$\{\mathfrak{x} \in [-\hat{\kappa}l, \hat{\kappa}l]^N \mid \mathfrak{x}_N \leq x_N\} \cap \{u = 0\} \neq \emptyset,$$

take $x^\star$ so that $(x^\star)' = x'$, $u(x^\star) = 0$ and $x_N^\star$ as low as possible (in particular, from (2.68), $x_N^\star \leq x_N$). Let also, as above, $y := (0, \ldots, 0, -l/2)$. Then, choosing

$$e := \frac{x^\star - y}{|x^\star - y|},$$

we have that $x$ and $x^\star$ here agree with $\mathfrak{x}(e)$ and $x^\star(e)$ constructed here above and that (2.65) is fulfilled provided $\hat{\kappa}$ is small enough. Thus, the desired result follows, in this case, from (2.67).

If, on the other hand,

(2.69) $$\{\mathfrak{x} \in [-\hat{\kappa}l, \hat{\kappa}l]^N \mid \mathfrak{x}_N \leq x_N\} \cap \{u = 0\} = \emptyset,$$

notice that $x_N + \theta + \kappa l \geq \theta + (\kappa - \hat{\kappa})l > 0$, provided that $\hat{\kappa} \leq \kappa$ and define

$$y^\star := x - (x_N + \theta + \kappa l)e_N$$

and
$$e := \frac{y^\star - y}{|y^\star - y|}.$$

Notice also that $y_N^\star + \kappa l = -\theta$, hence
$$B_{\kappa l}(y^\star) \subseteq [-\kappa l, \kappa l]^{N-1} \times (-\infty, -\theta] \subseteq \{u < 0\}.$$

Therefore, by the first item in Lemma 2.21, we have that $\Psi^{y+te,\kappa l}$ does not touch $u$ for $t \in [0, |y^\star - y|]$. In particular, for $t = |y^\star - y|$,
$$\begin{aligned} u(x) &\leq \Psi^{y+te,\kappa l}(x) = \\ &= \Psi^{y^\star,\kappa l}(x) = \\ &= g_{\kappa l}(x_N + \theta). \end{aligned}$$

This proves the desired result also in case (2.69) holds and it completes the proof of Lemma 2.23. □

COROLLARY 2.24. *Let $l > 0$, $\theta > 0$. Let $u$ be a local minimizer for $\mathcal{F}$ in*
$$\Omega := \left\{ (x', x_N) \in \mathbb{R}^{N-1} \times \mathbb{R} \,\middle|\, |x'| < l, \, |x_N| < l \right\}.$$
*Assume that $|u| \leq 1$, that $\theta/l$ is suitably small, that $u(0) = 0$, that $u(x) > 0$ if $x_N \geq \theta$ and that $u(x) < 0$ if $x_N \leq -\theta$. Then, there exists a suitable constant $c \in (0, 1]$ so that*
$$(-1 + s_{cl}, 1 - s_{cl}) \subseteq u(\Omega).$$

PROOF. By Lemma 2.23 and the fact that $g_l(t) = -1 + s_l$ if $t \leq -l/2$ (recall Lemma 2.20), we have that, if $x_N \leq -3\kappa l/4$,
$$u(x) \leq g_{\kappa l}(x_N + \theta) = -1 + s_{\kappa l},$$
provided that $\theta/l$ small enough. Analogously, by Lemma 2.23 and the fact that $\widetilde{g}_l(t) = 1 - s_l$ if $t \geq l/2$ (recall Lemma 2.22), we get that, if $x_N \geq 3\widetilde{\kappa}l/4$,
$$u(x) \geq \widetilde{g}_{\widetilde{\kappa}l}(x_N - \theta) = 1 - s_{\widetilde{\kappa}l}.$$

The inequalities above and the continuity of $u$ (see [15] or [34]) imply the desired result. □

In the following, we will often slide the barriers in a given direction. More precisely, we will start from a configuration in which the barrier is above a subsolution $u$ and then we slide the barrier until it touches the graph of $u$. With some poetry, we may think that the barrier is like a ship which moves forward until it touches the land $u$: of course, the ship will touch the land with the fore, not with the aft. Next result gives a formal justification of this fact:

LEMMA 2.25. *Let $u \in C(\Omega)$. Let $\xi = (\xi_1, \ldots, \xi_N) \in \mathbb{R}^N$ with $|\xi| = 1$ and let $\hat{\xi} := (\xi, 0) \in \mathbb{R}^{N+1}$. Assume that $g_{\mathbb{S}(Y-t\hat{\xi},R)} > u$ in their common domain of definition for any $t \in (0, 1]$. Assume also that $g_{\mathbb{S}(Y,R)}$ touches $u$ from above at some point $X = (x, x_{N+1})$, that is, assume that $g_{\mathbb{S}(Y,R)} \geq u$ and $g_{\mathbb{S}(Y,R)}(x) = u(x) = x_{N+1}$. Then,*
$$(x - y) \cdot \xi \geq 0.$$

PROOF. By construction,
$$g_{y_{N+1},R}\Big(|x-y| + H_0(y_{N+1}) - R\Big) =$$
$$= g_{\mathbb{S}(Y,R)}(x) =$$
$$= u(x) \le$$
$$\le g_{\mathbb{S}(Y-t\hat{\xi},R)}(x) =$$
$$= g_{y_{N+1},R}\Big(|x-y+t\xi| + H_0(y_{N+1}) - R\Big)$$
for any $t \in [0,1]$, from which
$$|x-y| \le |x-y+t\xi|$$
for any $t \in [0,1]$. This says that the function
$$f(t) := |x-y+t\xi|^2$$
attains its minimum in the domain $[0,1]$ at $t = 0$. Thus, $f'(0) \ge 0$, which gives the desired estimate. □

CHAPTER 3

# Geometry of the touching points

This section, which is very technical, follows many of the ideas in Chapter 4 of [**31**] (we provide full details for the reader's facility). The main result of this section will be Proposition 3.14, in which we investigate the measure theoretic properties of (an $N$-dimensional projection of) the set of possible touching points between a subsolution of (1.5) and the barrier $\mathbb{S}(Y, R)$ introduced here above. Roughly speaking, we will prove in Proposition 3.14 that the measure of the projection of the "first occurrence" touching points controls the measure of the projection of the centers of the corresponding surfaces.

For this scope, given

$$(3.1) \qquad \xi := (\xi_1, \ldots, \xi_N, 0) \in \mathbb{R}^{N+1},$$

with $|\xi| = 1$, we define $\mathfrak{P}_\xi$ as the hyperplane in $\mathbb{R}^{N+1}$ orthogonal to $\xi$, i.e.,

$$\mathfrak{P}_\xi := \{ X \in \mathbb{R}^{N+1} \mid \xi \cdot X = 0 \}$$

We also denote by $\pi_\xi$ the projection onto $\mathfrak{P}_\xi$, i.e.,

$$\pi_\xi(X) := X - (\xi \cdot X)\xi, \qquad \forall X \in \mathbb{R}^{N+1}.$$

With a slight abuse of notation, we will sometimes identify $\xi$ with its $N$-dimensional projection, implicitly dropping the zero in the last coordinate, that is, we will write

$$\xi := (\xi_1, \ldots, \xi_N) \in \mathbb{R}^N,$$

instead of (3.1).

Let now $u \in C^1([-C^\sharp l, C^\sharp l]^N)$ be a weak Sobolev subsolution of (1.5), with $|u| \leq 1$. Here we will fix $C^\sharp$ suitably large (also, $l$ and $R$ are fixed and suitably large, and $l/R$ is assumed conveniently small). Let us now define the set of first contact points as follows. Given a compact set[1] $\mathfrak{A} \subseteq \mathfrak{P}_\xi \subset \mathbb{R}^N$, we define

$$\begin{aligned}
\widetilde{\mathfrak{A}} = \widetilde{\mathfrak{A}}_{\mathfrak{A}} \ := \ & \{ Y \in \mathbb{R}^{N+1} \quad \text{s.t.} \\
& \exists \hat{Y} \in \mathfrak{A}, \, t_{\hat{Y}} \in \mathbb{R} \quad \text{s.t.} \quad Y = \hat{Y} + t_{\hat{Y}} \xi, \\
& g_{\mathbb{S}(\hat{Y}+t\xi, R)} > u \text{ for any } t < t_{\hat{Y}}, \\
& g_{\mathbb{S}(Y,R)} \geq u \\
& \text{and} \quad \exists x \in \mathbb{R}^{N+1} \quad \text{s.t.} \quad g_{\mathbb{S}(Y,R)}(x) = u(x) \}.
\end{aligned}$$

---

[1] The closure of $\mathfrak{A}$ will be used to deduce closure and measurability properties of sets of interest: see, e.g., Lemma 3.2 and Proposition 3.14 here below.

We refer to $\widetilde{\mathfrak{A}}$ as the set of the "centers". Moreover, we define

$$\widetilde{\mathfrak{B}} = \widetilde{\mathfrak{B}}_{\mathfrak{A}} := \{X = (x, x_{N+1}) \in \mathbb{R}^{N+1} \quad \text{s.t.} \quad \exists Y \in \widetilde{\mathfrak{A}} \quad \text{s.t.}$$
$$g_{\mathbb{S}(Y,R)} \geq u \quad \text{and}$$
$$g_{\mathbb{S}(Y,R)}(x) = u(x) = x_{N+1}\}$$

and we refer to $\widetilde{\mathfrak{B}}$ as the set of "first contact points". Roughly speaking, we are sliding our barriers until it touches the graph of $u$ for the first time: the set $\widetilde{\mathfrak{B}}$ collects all such first occurrence contact points, while $\widetilde{\mathfrak{A}}$ collects the corresponding centers. In this section, we will assume that

(3.2) $$\mathfrak{A} \subseteq [-C^{\sharp}l/2, C^{\sharp}l/2]^N \times [-1/4, 1/4]$$

and that

(3.3) $$\left\{X = (x, x_{N+1}) \in \mathbb{R}^{N+1} \text{ s.t. } \exists Y \in \mathbb{R}^{N+1} \text{ s.t. } \pi_\xi Y \in \mathfrak{A}, \; g_{\mathbb{S}(Y,R)} \geq u \text{ and} \atop g_{\mathbb{S}(Y,R)}(x) = u(x)\right\} \cap \partial[-C^{\sharp}l, C^{\sharp}l]^N = \emptyset.$$

We also define

$$\mathfrak{B} := \pi_\xi(\widetilde{\mathfrak{B}}).$$

We also denote the graph of $u$ by $\mathfrak{G}$, that is we set

(3.4) $$\mathfrak{G} := \{x_{N+1} = u(x)\}.$$

We now show some properties of the above defined sets. First of all, from (3.3) and Corollary 2.14, we have that:

LEMMA 3.1.
$$\widetilde{\mathfrak{B}} \subseteq \{X \in \mathbb{R}^{N+1} \mid |x_{N+1}| < 1/2\}.$$

We now show that the compactness property of $\mathfrak{A}$ is inherited by the other sets defined above:

LEMMA 3.2. *If $l/R$ is small enough, then $\widetilde{\mathfrak{A}}$, $\mathfrak{B}$ and $\widetilde{\mathfrak{B}}$ are compact sets.*

PROOF. Note that $\widetilde{\mathfrak{B}} \in [-C^{\sharp}l, C^{\sharp}l]^N \times [-1,1]$, hence $\widetilde{\mathfrak{B}}$ is bounded. Therefore, $\mathfrak{B}$ and $\widetilde{\mathfrak{A}}$ are also bounded. Thence, we only need to show that the above sets are closed. Let us first show that $\widetilde{\mathfrak{A}}$ is closed. For this, let $Y_k \in \widetilde{\mathfrak{A}}$ converge to some $Y_\infty$. We need to show that $Y_\infty \in \widetilde{\mathfrak{A}}$.

For this scope, note that, since $\mathfrak{A}$ is closed and $\pi_\xi Y_k \in \pi_\xi(\widetilde{\mathfrak{A}}) \subseteq \mathfrak{A}$, we have that $\pi_\xi Y_\infty \in \mathfrak{A}$. Also, since $Y_k \in \widetilde{\mathfrak{A}}$, we have that there exists $X_k \in [-C^{\sharp}l, C^{\sharp}l]^N \times [-1,1]$ so that

(3.5) $$\begin{cases} g_{\mathbb{S}(Y_k,R)}(x_k) = u(x_k) \\ g_{\mathbb{S}(Y_k,R)} \geq u \quad \text{and} \\ g_{\mathbb{S}(Y_k - t\xi,R)} > u, \end{cases}$$

for any $t > 0$. Of course, up to subsequence, we may assume that $X_k$ converges to some point $X_\infty$. Thus, passing to the limit (3.5), we obtain that

$$\begin{cases} g_{\mathbb{S}(Y_\infty,R)}(x_\infty) = u(x_\infty) \\ g_{\mathbb{S}(Y_\infty,R)} \geq u \quad \text{and} \\ g_{\mathbb{S}(Y_\infty - t\xi,R)} \geq u, \end{cases}$$

## 3. GEOMETRY OF THE TOUCHING POINTS

for any $t > 0$. Hence, to show that $Y_\infty \in \widetilde{\mathfrak{A}}$, we need to prove the strict inequality in the last relation here above, i.e., we need to show that

(3.6) $$g_{\mathbb{S}(Y_\infty - t\xi, R)} > u$$

for any $t > 0$. We argue by contradiction: assume that there exists $x_\sharp$ so that $|x_\sharp| \leq \text{const } l$ and $t_\sharp > 0$, such that

$$g_{\mathbb{S}(Y_\infty - t_\sharp \xi, R)}(x_\sharp) = u(x_\sharp).$$

Note that, by (3.3) and Corollary 2.14, $|g_{\mathbb{S}(Y_\infty, R)}(x_\infty)| \leq 1/2$, thus

(3.7) $$|x_\infty - y_\infty| \geq \text{const } R.$$

Also, by (3.2), we get that

$$\begin{aligned}|\pi_\xi(x_\infty - y_\infty)| &\leq |x_\infty| + |\pi_\xi y_\infty| = \\ &= |x_\infty| + |\pi_{e_N}(\pi_\xi Y_\infty)| \leq \\ &\leq \text{const } l.\end{aligned}$$

Thus, from (3.7), we have that

(3.8) $$\angle(x_\infty - y_\infty, \xi) \leq \text{const } l/R.$$

Furthermore, from Lemma 2.25,

$$(x_\infty - y_\infty) \cdot \xi \geq 0.$$

This, (3.8) and (3.7) say that

$$\begin{aligned}(x_\infty - y_\infty) \cdot \xi &= |(x_\infty - y_\infty) \cdot \xi| = \\ &= |x_\infty - y_\infty| \cos\angle(x_\infty - y_\infty, \xi) \geq \\ &\geq \text{const } R,\end{aligned}$$

provided that $l/R$ is small enough. For this reason,

(3.9) $$\begin{aligned}(x_\sharp - y_\infty) \cdot \xi &\geq (x_\infty - y_\infty) \cdot \xi - |x_\infty| - |x_\sharp| \geq \\ &\geq \text{const } R - \text{const } l \geq \\ &\geq 0,\end{aligned}$$

if $l/R$ is small enough. Thus, from (3.9), we deduce that

$$\begin{aligned}|x_\sharp - y_\infty + t_\sharp \xi|^2 &= |x_\sharp - y_\infty|^2 + t_\sharp^2 + 2t_\sharp \xi \cdot (x_\sharp - y_\infty) > \\ &> |x_\sharp - y_\infty|^2.\end{aligned}$$

As a result, we infer that

$$\begin{aligned}u(x_\sharp) &= g_{\mathbb{S}(Y_\infty - t_\sharp \xi, R)}(x_\sharp) = \\ &= g_{y_{\infty, N+1}, R}\Big(|x_\sharp - y_\infty + t_\sharp \xi| + H_0(y_{\infty, N+1}) - R\Big) > \\ &> g_{y_{\infty, N+1}, R}\Big(|x_\sharp - y_\infty| + H_0(y_{\infty, N+1}) - R\Big) = \\ &= g_{\mathbb{S}(Y_\infty, R)}(x_\sharp) \geq \\ &\geq u(x_\sharp).\end{aligned}$$

This contradiction proves (3.6). Hence, $Y_\infty \in \widetilde{\mathfrak{A}}$ and therefore $\widetilde{\mathfrak{A}}$ is closed.

Note now that once we know that $\widetilde{\mathfrak{B}}$ is closed, it easily follows that $\mathfrak{B}$ is also closed. Thus, to end the proof of this result, we need to prove that $\widetilde{\mathfrak{B}}$ is closed. For this, let us consider a sequence $X_k \in \widetilde{\mathfrak{B}}$ so that
$$\lim_{k \to +\infty} X_k = X_\infty.$$
Our aim is to show that $X_\infty \in \widetilde{\mathfrak{B}}$. For this, observe that, since $X_k \in \widetilde{\mathfrak{B}}$, there exists $Y_k \in \widetilde{\mathfrak{A}}$ such that $g_{\mathbb{S}(Y_k,R)} \geq u$ and $g_{\mathbb{S}(Y_k,R)}(x_k) = u(x_k) = x_{k,N+1}$. Since we proved that $\widetilde{\mathfrak{A}}$ is compact, possibly taking subsequences, we may assume that $Y_k$ tends to $Y_\infty \in \widetilde{\mathfrak{A}}$. Thus, passing to the limit here above we deduce that $g_{\mathbb{S}(Y_\infty,R)} \geq u$ and $g_{\mathbb{S}(Y_\infty,R)}(x_\infty) = u(x_\infty) = x_{\infty,N+1}$, with $Y_\infty \in \widetilde{\mathfrak{A}}$. This proves that $X_\infty \in \widetilde{\mathfrak{B}}$, thence $\widetilde{\mathfrak{B}}$ is closed. $\square$

LEMMA 3.3. *For any $X \in \mathbb{R}^{N+1}$ with $X \in \widetilde{\mathbb{S}}(Y,R)$ and $\nu \in S^N$, let*[2]

$$\nu^{\widetilde{\mathbb{S}}(Y,R)}(X) := \frac{\left(-\nabla g_{\widetilde{\mathbb{S}}(Y,R)}(x), 1\right)}{\sqrt{1 + |\nabla g_{\widetilde{\mathbb{S}}(Y,R)}(x)|^2}} \in \mathbb{R}^{N+1},$$

$$\omega(X,\nu) := \frac{R}{\overline{C_0}} \left( \frac{\nu_{N+1}}{|(\nu_1,\ldots,\nu_N)|} - H_0'(x_{N+1}) \right) \in \mathbb{R},$$

$$\sigma(X,\nu) := -\frac{\overline{C_0}}{2R} \omega^2(X,\nu) + H_0(x_{N+1}) - H_0(x_{N+1} + \omega(X,\nu)) + R \in \mathbb{R},$$

$$F(X,\nu) := \left( x + \frac{(\nu_1,\ldots,\nu_N)}{|(\nu_1,\ldots,\nu_N)|} \sigma(X,\nu),\ x_{N+1} + \omega(X,\nu) \right) \in \mathbb{R}^{N+1}.$$

*Then,*

(3.10) $$Y = F(X, \nu^{\widetilde{\mathbb{S}}(Y,R)}(X)),$$

*for any $X \in \widetilde{\mathbb{S}}(Y,R)$.*

PROOF. For short, we will set here $\widetilde{\mathbb{S}} := \widetilde{\mathbb{S}}(Y,R)$. Recalling Lemma B.11, we see that there exists $\sigma \in \mathbb{R}$ so that
$$x - y = \sigma \frac{(\nu_1^{\widetilde{\mathbb{S}}}(X), \ldots, \nu_N^{\widetilde{\mathbb{S}}}(X))}{|(\nu_1^{\widetilde{\mathbb{S}}}(X), \ldots, \nu_N^{\widetilde{\mathbb{S}}}(X))|}.$$
Using that $x_{N+1} = g_{\widetilde{\mathbb{S}}(Y,R)}(x)$, (2.31), (2.33) and (2.35), we also gather that
$$H_0(y_{N+1}) + |x-y| - R = H_0(x_{N+1}) - \frac{\overline{C_0}}{2R}(x_{N+1} - y_{N+1})^2.$$
Hence, if $\omega := y_{N+1} - x_{N+1}$, we gather
$$\sigma = |x-y| = H_0(x_{N+1}) - H_0(x_{N+1} + \omega) + R - \frac{\overline{C_0}}{2R}\omega^2.$$
This determines $\sigma$, as desired, we now need to determine $\omega$. Using (2.2) (with $s_0 := y_{N+1}$ and $s := x_{N+1}$) and Definition 2.5 (with the fact that $|s| < 1/2$ as

---

[2] Of course, $\nu^{\widetilde{\mathbb{S}}(Y,R)}(X)$ is normal to $\widetilde{\mathbb{S}}(Y,R)$ at the point $X$.

pointed out above), we get

$$\begin{aligned}\omega &= -(x_{N+1} - y_{N+1}) = \\ &= \frac{R}{\overline{C}_0}\left(\frac{1}{\left(\frac{p}{p-1}\varphi(x_{N+1})\right)^{1/p}} - \frac{1}{\left(\frac{p}{p-1}h_0(x_{N+1})\right)^{1/p}}\right) = \\ &= \frac{R}{\overline{C}_0}\left(\frac{1}{\left(\frac{p}{p-1}h_{s_0,R}(x_{N+1})\right)^{1/p}} - \frac{1}{\left(\frac{p}{p-1}h_0(x_{N+1})\right)^{1/p}}\right).\end{aligned}$$

Thus, from Definition 2.12 and (2.42), we have that

$$\begin{aligned}\omega &= \frac{R}{\overline{C}_0}\left(H'_{s_0,R}(x_{N+1}) - H'_0(x_{N+1})\right) = \\ &= \frac{R}{\overline{C}_0}\left(\frac{1}{|\nabla g_{\widetilde{\mathbb{S}}(Y,R)}(x)|} - H'_0(x_{N+1})\right),\end{aligned}$$

and, therefore,

$$\omega = \frac{R}{\overline{C}_0}\left(\frac{\nu^{\widetilde{\mathbb{S}}}_{N+1}(X)}{|(\nu^{\widetilde{\mathbb{S}}}_1(X),\ldots,\nu^{\widetilde{\mathbb{S}}}_N(X))|} - H'_0(x_{N+1})\right),$$

which determines $\omega$. $\square$

COROLLARY 3.4. *Let the notation of Lemma 3.3 holds. For $X = (x, x_{N+1}) \in \mathbb{R}^N$, with $x_{N+1} = u(x)$, let*[3]

$$(3.11) \qquad \nu^u(X) := \frac{\left(-\nabla u(x), 1\right)}{\sqrt{1 + |\nabla u(x)|^2}} \in \mathbb{R}^{N+1},$$

*Let $X, Y \in \mathbb{R}^N$ be so that*

$$(3.12) \qquad \begin{aligned} g_{\mathbb{S}(Y,R)}(z) &\geq u(z) \quad \forall z \in \mathbb{R}^N, \\ g_{\mathbb{S}(Y,R)}(x) &= u(x) = x_{N+1}.\end{aligned}$$

*Then, $Y = F(X, \nu^u(X))$.*

PROOF. By (3.12), we have that the graph of $u$ and the surface $\mathbb{S}(Y,R)$ are tangent at the point $X$, therefore

$$\nu^u(X) = \nu^{\mathbb{S}(Y,R)}(X),$$

hence the claim follows from Lemma 3.3. $\square$

LEMMA 3.5. *In the notation of Corollary 3.4, if $X \in \widetilde{\mathfrak{B}}$, then there exists a positive constant $c$ so that*

$$c \leq \nu^u_{N+1}(X) \leq 1 - c.$$

---

[3]Note that $\nu^u(X)$ is a unit vector, normal to the graph of $u$ at the point $X = (x, u(x))$.

PROOF. Since, by Lemma 3.1, $|x_{N+1}| \leq 1/2$,

$$|\nabla u(x)| = |\nabla g_{\mathbb{S}(Y,R)}(x)| = \left|\left(\frac{p}{p-1}h_{s_0,R}(x_{N+1})\right)^{1/p}\right| \in [1/C, C],$$

for a suitable constant $C$, which implies the desired claim. □

LEMMA 3.6. *Let the notation of Lemma 3.3 and Corollary 3.4 hold. Let*

$$\mathcal{Y}(X) := F(X, \nu^u(X))$$

*and let $D_X \mathcal{Y}$ be the differential map. Then, there exists a positive constant $C$ such that*

$$|D_X \mathcal{Y}(X)| \leq C,$$

*for any $X \in \widetilde{\mathfrak{B}}$.*

PROOF. By direct inspection,

(3.13) $\quad D_X \mathcal{Y}(X) = D_X F(X, \nu^u(X)) + D_\nu F(X, \nu^u(X)) D_X \nu^u(X).$

On the other hand, by differentiating (3.10),

(3.14) $\quad 0 = D_X F(X, \nu^{\mathbb{S}(Y,R)}(X)) + D_\nu F(X, \nu^{\mathbb{S}(Y,R)}(X)) D_X \nu^{\mathbb{S}(Y,R)}(X).$

Moreover, if $X \in \widetilde{\mathfrak{B}}$, then

(3.15) $\quad \nabla g_{\widetilde{\mathbb{S}}(Y,R)}(x) = \nabla u(x),$

and so

$$\nu^u(X) = \nu^{\widetilde{\mathbb{S}}(Y,R)}(X).$$

Thus, from (3.13) and (3.14), we gather that

$$D_X \mathcal{Y}(X) = D_\nu F(X, \nu^{\widetilde{\mathbb{S}}(Y,R)}(X)) \left(D_X \nu^u(X) - D_X \nu^{\widetilde{\mathbb{S}}(Y,R)}(X)\right),$$

for any $X \in \widetilde{\mathfrak{B}}$. By the definitions given in Lemma 3.3, one sees that

$$|D_\nu F(X, \nu^{\widetilde{\mathbb{S}}(Y,R)}(X))| \leq \text{const } R,$$

and so we get from the above relation that

$$|D_X \mathcal{Y}(X)| \leq \text{const } R \left|D_X \nu^u(X) - D_X \nu^{\widetilde{\mathbb{S}}(Y,R)}(X)\right|.$$

Therefore, in the light of (3.15), Lemma B.12, Remark B.13 and Lemma 3.5,

$$|D_X \mathcal{Y}(X)| \leq \text{const } R \left|D^2 u(x) - D^2 g_{\widetilde{\mathbb{S}}(Y,R)}(x)\right|.$$

Also, since $g_{\widetilde{\mathbb{S}}(Y,R)}$ touches $u$ from above at $X$, we have that

(3.16) $\quad D^2 g_{\widetilde{\mathbb{S}}(Y,R)}(x) - D^2 u(x)$ is a non-negative definite matrix

and therefore

$$\text{const } \left|D^2 u(x) - D^2 g_{\widetilde{\mathbb{S}}(Y,R)}(x)\right| \leq \Delta \left(g_{\widetilde{\mathbb{S}}(Y,R)} - u\right)(x).$$

Thence, we have obtained the following estimate:

(3.17) $\quad |D_X \mathcal{Y}(X)| \leq \text{const } R \Delta \left(g_{\widetilde{\mathbb{S}}(Y,R)} - u\right)(x).$

Notice now that $u$ is $C^2$ near $X$ by standard elliptic results, since $\nabla u(x) \neq 0$ thanks to (3.15). Thence, making use of (3.15), we get

$$
\begin{aligned}
\Delta_p g_{\widetilde{\mathbb{S}}(Y,R)}(x) - \Delta_p u(x) = \\
= \ & |\nabla g_{\widetilde{\mathbb{S}}(Y,R)}(x)|^{p-2} \left( \Delta g_{\widetilde{\mathbb{S}}(Y,R)}(x) - \Delta u(x) \right) + \\
& + (p-2) |\nabla g_{\widetilde{\mathbb{S}}(Y,R)}(x)|^{p-4} \nabla g_{\widetilde{\mathbb{S}}(Y,R)}(x) \cdot \\
& \cdot \left[ \left( D^2 g_{\widetilde{\mathbb{S}}(Y,R)}(x) - D^2 u(x) \right) \cdot \nabla g_{\widetilde{\mathbb{S}}(Y,R)}(x) \right].
\end{aligned}
\tag{3.18}
$$

We need now to distinguish two cases. If $p \geq 2$, we use Lemma 3.5 and (3.16) in order to deduce from (3.18) that

$$
\begin{aligned}
\Delta_p g_{\widetilde{\mathbb{S}}(Y,R)}(x) - \Delta_p u(x) \geq & \\
\geq \ & |\nabla g_{\widetilde{\mathbb{S}}(Y,R)}(x)|^{p-2} \left( \Delta g_{\widetilde{\mathbb{S}}(Y,R)}(x) - \Delta u(x) \right) \geq \\
\geq \ & \operatorname{const} \left( \Delta g_{\widetilde{\mathbb{S}}(Y,R)}(x) - \Delta u(x) \right).
\end{aligned}
\tag{3.19}
$$

On the other hand, if $1 < p < 2$, (3.18), Lemma 3.5 and (3.16) give that

$$
\begin{aligned}
\Delta_p g_{\widetilde{\mathbb{S}}(Y,R)}(x) - \Delta_p u(x) \geq & \\
\geq \ & |\nabla g_{\widetilde{\mathbb{S}}(Y,R)}(x)|^{p-2} \left( \Delta g_{\widetilde{\mathbb{S}}(Y,R)}(x) - \Delta u(x) \right) - \\
& - (2-p) |\nabla g_{\widetilde{\mathbb{S}}(Y,R)}(x)|^{p-2} |D^2 g_{\widetilde{\mathbb{S}}(Y,R)}(x) - D^2 u(x)| \geq \\
\geq \ & |\nabla g_{\widetilde{\mathbb{S}}(Y,R)}(x)|^{p-2} [1 - (2-p)] \left( \Delta g_{\widetilde{\mathbb{S}}(Y,R)}(x) - \Delta u(x) \right) \geq \\
\geq \ & \operatorname{const}(p-1) \left( \Delta g_{\widetilde{\mathbb{S}}(Y,R)}(x) - \Delta u(x) \right).
\end{aligned}
\tag{3.20}
$$

In any case, for any $p \in (1, +\infty)$, (3.19) and (3.20) give that

$$
\begin{aligned}
\Delta_p g_{\widetilde{\mathbb{S}}(Y,R)}(x) - \Delta_p u(x) & \\
\geq \ & \operatorname{const} \left( \Delta g_{\widetilde{\mathbb{S}}(Y,R)}(x) - \Delta u(x) \right).
\end{aligned}
\tag{3.21}
$$

Furthermore, exploiting Proposition 2.18 and the fact that $u$ is a subsolution of (1.5), we have that

$$
\Delta_p g_{\widetilde{\mathbb{S}}(Y,R)}(x) - \Delta_p u(x) \leq \frac{C}{R}.
\tag{3.22}
$$

The desired result now follows from (3.17), (3.21) and (3.22). □

Recalling (3.4), we define
$$\mathfrak{S} := \mathcal{Y}(\mathfrak{G}).$$

The construction of $\widetilde{\mathfrak{A}}$ and $\widetilde{\mathfrak{B}}$ easily implies the following observation:

LEMMA 3.7. $\widetilde{\mathfrak{A}}$ and $\widetilde{\mathfrak{B}}$ belong to Lipschitz surfaces. More precisely, $\widetilde{\mathfrak{B}}$ lies in $\mathfrak{G}$ while $\widetilde{\mathfrak{A}}$ lies in $\mathfrak{S}$.

The next observation will say that $\widetilde{\mathfrak{A}}$ and $\widetilde{\mathfrak{B}}$ are graphs with respect to the $\xi$-direction:

LEMMA 3.8. $\pi_\xi$ is injective on $\widetilde{\mathfrak{A}}$ and on $\widetilde{\mathfrak{B}}$.

PROOF. $\pi_\xi$ in injective on $\widetilde{\mathfrak{A}}$ by construction. Let us show that is also injective on $\widetilde{\mathfrak{B}}$. Assume, by contradiction, that $X^{(1)} \in \widetilde{\mathfrak{B}}$ and $X^{(2)} = X^{(1)} + \tau\xi \in \widetilde{\mathfrak{B}}$, with $\tau > 0$. Let also $Y^{(1)}$ and $Y^{(2)}$ be the corresponding centers in $\widetilde{\mathfrak{A}}$, i.e., for $i = 1, 2$, let $Y^{(i)} \in \widetilde{\mathfrak{A}}$ be so that $g_{\mathbb{S}(Y^{(i)},R)}$ touches $u$ for the first time at $X^{(i)}$.

Since $\xi_{N+1} = 0$,
$$x^{(1)}_{N+1} = x^{(2)}_{N+1},$$
therefore,

(3.23)
$$g_{\mathbb{S}(Y^{(1)},R)}(x^{(1)}) = x^{(1)}_{N+1} = u(x^{(1)}) =$$
$$= x^{(2)}_{N+1} = g_{\mathbb{S}(Y^{(2)},R)}(x^{(2)}) = u(x^{(2)}).$$

If now we consider
$$\bar{Y} := Y^{(2)} - \tau\xi$$
it follows that

(3.24) $\quad g_{\mathbb{S}(\bar{Y},R)}(x^{(1)}) = g_{\mathbb{S}(Y^{(2)}-\tau\xi,R)}(x^{(2)} - \tau\xi) = g_{\mathbb{S}(Y^{(2)},R)}(x^{(2)}) = u(x^{(1)}).$

On the other hand, since $Y^{(2)} \in \widetilde{\mathfrak{A}}$, $g_{\mathbb{S}(\bar{Y},R)} > u$, in contradiction with (3.24). □

Given $X \in \mathbb{R}^{N+1}$, we now define
(3.25) $$\Sigma_X := \{Z \mid X \in \widetilde{\mathbb{S}}(Z, R)\}.$$
In other words, given a point $X$, $\Sigma_X$ is the surface containing all the centers of the surfaces $\widetilde{\mathbb{S}}(\cdot, R)$ to which $X$ belongs. Let us now investigate the properties of $\Sigma_X$:

LEMMA 3.9. *Let $\mathcal{Y}$ be as in Lemma 3.6. Then, $\mathcal{Y}(X) \in \Sigma_X$ for any $X \in \widetilde{\mathfrak{B}}$.*

PROOF. Since $X \in \widetilde{\mathfrak{B}}$, there exists $Y$ so that (3.12) holds. Then, by Corollary 3.4,
$$\Sigma_X \ni Y = F(X, \nu^u(X)) = \mathcal{Y}(X).$$
□

We denote by $\nu^{\Sigma_X}(Z)$ the unit normal vector to the surface $\Sigma_X$ at a point $Z \in \Sigma_X$ (in a fixed orientation). Such definition is well posed, since $\Sigma_X$ is a Lipschitz surface, as we show here below. Also, we can express $\nu^{\Sigma_X}$ in terms of the normal to $\widetilde{\mathbb{S}}(Y, R)$, according to the following result:

LEMMA 3.10. *Let $\mathcal{Y}$ be as in Lemma 3.6 and let $X \in \widetilde{\mathfrak{B}}$. Then $\Sigma_X$ is a Lipschitz rotation surface with axis parallel to $e_{N+1}$ and passing through $X$. Also, $\nu^{\Sigma_X}(\mathcal{Y}(X))$ belongs to the space spanned by $e_{N+1}$ and $\nu^{\widetilde{\mathbb{S}}(\mathcal{Y}(X),R)}(X)$.*

PROOF. Let us first show that $\Sigma_X$ is a Lipschitz rotation surface. By (2.35), we have that $Z \in \Sigma_X$ if and only if
$$x_{N+1} = g_{\widetilde{\mathbb{S}}(Z,R)}(x) = \rho_{z_{N+1},R}(H_0(z_{N+1}) + |z - x| - R),$$
that is, by (2.33) and (2.31), if and only if
$$H_0(x_{N+1}) - \frac{\overline{C_0}}{2R}(x_{N+1} - z_{N+1})^2 = H_0(z_{N+1}) + |z - x| - R.$$
We now define
$$\mathfrak{H}_{x_{N+1},R}(\zeta) := H_0(\zeta) + \frac{\overline{C_0}}{2R}(x_{N+1} - \zeta)^2.$$

Then, $\mathfrak{H}$ is strictly increasing in $[-1/2, 1/2]$; thus $Z \in \Sigma_X$ if and only if
$$z_{N+1} = \mathfrak{H}^{-1}_{x_{N+1},R}\Big(H_0(x_{N+1}) + R - |z - x|\Big).$$
This proves the rotational symmetry and the Lipschtiz properties of $\Sigma_X$.

Consequently, by Lemma B.10, we gather that $\nu^{\Sigma_X}(\mathcal{Y}(X))$ is in the space spanned by $(\pi_{e_{N+1}}\mathcal{Y}(X) - x)$ and $e_{N+1}$. But $\nu^{\widetilde{\mathbb{S}}(\mathcal{Y}(X),R)}(X)$ is also in the space spanned by $(\pi_{e_{N+1}}\mathcal{Y}(X) - x)$ and $e_{N+1}$, as follows by the radial symmetry of $\widetilde{\mathbb{S}}(\mathcal{Y}(X), R)$ and Lemma B.10 again; moreover, $\nu^{\widetilde{\mathbb{S}}(\mathcal{Y}(X),R)}(X)$ is not parallel to $e_{N+1}$ (because $|x_{N+1}| \leq 1/2$ due to Lemma 3.1 and so $\nabla g_{\widetilde{\mathbb{S}}(\mathcal{Y}(X),R)}(x) \neq 0$). Therefore, $\nu^{\Sigma_X}(\mathcal{Y}(X))$ belongs to the space spanned by $\nu^{\widetilde{\mathbb{S}}(\mathcal{Y}(X),R)}(X)$ and $e_{N+1}$. □

LEMMA 3.11. *There exists a positive constant $C$ such that*
$$|\nu^{\Sigma_X}(\mathcal{Y}(X)) \cdot \xi| \leq C |\nu^u(X) \cdot \xi|,$$
*for any $X \in \widetilde{\mathfrak{B}}$.*

PROOF. From Lemma 3.10, for any $X \in \widetilde{\mathfrak{B}}$, we have that
$$(3.26) \qquad \nu^{\Sigma_X}(\mathcal{Y}(X)) = \alpha\hat{\nu} + \beta e_{N+1},$$
for some $\alpha = \alpha(X)$ and $\beta = \beta(X) \in \mathbb{R}$, where we denoted
$$\hat{\nu} := \Big(\nu_1^{\widetilde{\mathbb{S}}(\mathcal{Y}(X),R)}(X), \ldots, \nu_N^{\widetilde{\mathbb{S}}(\mathcal{Y}(X),R)}(X), 0\Big).$$
Obviously, since $X \in \widetilde{\mathfrak{B}}$,
$$\hat{\nu} = \Big(\nu_1^u(X), \ldots, \nu_N^u(X), 0\Big).$$
Thence, by exploiting Lemma 3.5, we see that
$$|\hat{\nu}|^2 = 1 - |\nu_{N+1}^{\widetilde{\mathbb{S}}(\mathcal{Y}(X),R)}(X)|^2 \geq 1 - (1-c)^2 \geq c.$$
Thus, (3.26) implies that
$$\begin{aligned} c|\alpha| &\leq |\alpha\hat{\nu} \cdot \hat{\nu}| = \\ &= \Big|\big(\nu^{\Sigma_X}(\mathcal{Y}(X)) - \beta e_{N+1}\big) \cdot \hat{\nu}\Big| = \\ &= \Big|\nu^{\Sigma_X}(\mathcal{Y}(X)) \cdot \hat{\nu}\Big| \leq \\ &\leq 1, \end{aligned}$$
that is
$$|\alpha| \leq \text{const}.$$
Hence, being $\xi_{N+1} = 0$, (3.26) gives that
$$|\nu^{\Sigma_X}(\mathcal{Y}(X)) \cdot \xi| = |\alpha\hat{\nu} \cdot \xi| \leq \text{const}\,|\hat{\nu} \cdot \xi| = \text{const}\,|\nu^u \cdot \xi|.$$
□

LEMMA 3.12. *Let $X \in \mathfrak{G}$, with $|x_{N+1}| \leq 1/2$. Assume that $g_{\widetilde{\mathbb{S}}(Y,R)} \geq u$. Then, $Y$ is above $\Sigma_X$ (with respect to the $e_{N+1}$-direction).*

PROOF. Assume that $(y, y_{N+1}^*) \in \Sigma_X$. We need to show that $y_{N+1}^* \leq y_{N+1}$. For this purpose, note that, by construction, $X \in \widetilde{\mathbb{S}}((y, y_{N+1}^*), R)$, which says that

$$g_{\widetilde{\mathbb{S}}((y,y_{N+1}^*),R)}(x) = x_{N+1} = u(x) \leq g_{\widetilde{\mathbb{S}}(Y,R)}(x). \tag{3.27}$$

Fix now $R$, $x$ and $y$. For $|\zeta| \leq 1/2$, we define

$$f(\zeta) := g_{\widetilde{\mathbb{S}}((y,\zeta),R)}(x).$$

Recalling (2.35) and (2.31), we have that

$$H_0(f(\zeta)) - \frac{\overline{C_0}}{2R}(\zeta - f(\zeta))^2 = H_0(\zeta) + |x - y| - R.$$

Differentiating with respect to $\zeta$, we get that

$$H_0'(f(\zeta)) f'(\zeta) \geq H_0'(\zeta) - \frac{\text{const}}{R}.$$

In particular, since by definition $|f(\zeta)| \leq 1/2$, we get that $f'(\zeta) > 0$ if $R$ is large enough, thence $f$ is increasing. Since, by (3.27), we have that

$$f(y_{N+1}^*) \leq f(y_{N+1}),$$

we deduce that $y_{N+1}^* \leq y_{N+1}$, as desired. $\square$

LEMMA 3.13. *Let $X \in \widetilde{\mathfrak{B}}$. Then $\widetilde{\mathfrak{A}}$ touches $\Sigma_X$ from above at $\mathcal{Y}(X)$.*

PROOF. First, note that

$$\mathcal{Y}(X) \in \mathcal{Y}(\widetilde{\mathfrak{B}}) \subseteq \widetilde{\mathfrak{A}}.$$

On the other hand, $\mathcal{Y}(X) \in \Sigma_X$ by Lemma 3.9. Thus, to end the proof of this result we need to show that $\widetilde{\mathfrak{A}}$ is above $\Sigma_X$ (with respect to the $e_{N+1}$-direction). For this purpose, take $Y \in \widetilde{\mathfrak{A}}$. Then, by construction, $g_{\widetilde{\mathbb{S}}(Y,R)} \geq u$ (and equality holds at some point). Thus, by Lemma 3.12, $Y$ is above $\Sigma_X$. $\square$

The following is the main result of this section, in which a measure estimate for contact points is given:

PROPOSITION 3.14. *Let $1/c \leq l \leq cR$, for a suitably small positive constant $c$. Assume (3.2) and (3.3). Assume also that $\mathfrak{A}$ is the closure of an open set and that*

$$\begin{aligned} &\text{for any } Y \in \mathfrak{A} \text{ there exist } t \in \mathbb{R} \text{ and } x \in \mathbb{R}^N, \text{ such that} \\ &g_{\mathbb{S}(Y+t\xi,R)}(x) \leq u(x). \end{aligned} \tag{3.28}$$

*Then, denoting the $N$-dimensional Lebesgue measure by $\mathfrak{L}^N$, we have that*

$$\mathfrak{L}^N(\mathfrak{A}) \leq C\, \mathfrak{L}^N(\mathfrak{B}),$$

*for a suitable positive universal constant $C$.*

PROOF. Note that $\mathfrak{A}$ is closed by hypothesis and so is $\mathfrak{B}$ thanks to Lemma 3.2; in particular, $\mathfrak{A}$ and $\mathfrak{B}$ are measurable sets. Also, by (3.28), $\mathfrak{A} = \pi_\xi(\widetilde{\mathfrak{A}})$. What is more, since $\mathfrak{A}$ is the closure of an open set, Lemma 3.7 and Lemma 3.8 say that $\widetilde{\mathfrak{A}}$ is a Lipschitz surface which is also a continuous[4] graph over $\mathfrak{A}$ in the $\xi$-direction.

---

[4] The continuity of $(\pi_\xi\big|_{\widetilde{\mathfrak{A}}})^{-1}$ follows from the following elementary property: if $f: K \longrightarrow f(K)$ is continuous and injective and $K$ is compact, then $f^{-1}$ is continuous.

In particular, the normal $\nu^{\widetilde{\mathfrak{A}}}$ is well defined. Thanks to Lemma 3.13, we also have that
$$\nu^{\Sigma_X}(\mathcal{Y}(X)) = \nu^{\widetilde{\mathfrak{A}}}(\mathcal{Y}(X))$$
for any $X \in \widetilde{\mathfrak{B}}$, provided that $\pi_\xi(\mathcal{Y}(X))$ lies in the interior of $\mathfrak{A}$ (up to an orientation choice). Therefore, by Lemma 3.11,

(3.29) $$|\nu^{\widetilde{\mathfrak{A}}}(\mathcal{Y}(X)) \cdot \xi| \leq \text{const} \, |\nu^u(X) \cdot \xi|,$$

where $\nu^u(X)$ is the normal to $\mathfrak{G}$ at $X$, for any $X \in \widetilde{\mathfrak{B}}$, provided that $\pi_\xi(\mathcal{Y}(X))$ lies in the interior of $\mathfrak{A}$. We denote by $\widetilde{\mathfrak{B}}'$ this set, that is
$$\widetilde{\mathfrak{B}}' := \left\{ X \in \widetilde{\mathfrak{B}}, \text{ with } \pi_\xi(\mathcal{Y}(X)) \text{ in the interior of } \mathfrak{A} \right\}.$$

Also, applying the divergence theorem to (the interior of) $\widetilde{\mathfrak{A}}$, we have that

(3.30) $$\mathcal{L}^N(\mathfrak{A}) \leq \int_{\widetilde{\mathfrak{A}}} |\nu^{\widetilde{\mathfrak{A}}} \cdot \xi|,$$

where the above is a surface integral.

For $\varepsilon > 0$, let us now define
$$\widetilde{\mathfrak{B}}_\varepsilon := \bigcup_{X \in \widetilde{\mathfrak{B}}'} B_\varepsilon(Y) \cap \mathfrak{G},$$
where $\mathfrak{G}$, as in (3.4), denotes the graph of $u$. Then, $\widetilde{\mathfrak{B}}_\varepsilon$ is a Lipschitz surface contained in $\mathfrak{G}$. By the divergence theorem,
$$\mathcal{L}^N\left(\pi_\xi(\widetilde{\mathfrak{B}}_\varepsilon)\right) = \int_{\widetilde{\mathfrak{B}}_\varepsilon} \nu^{\widetilde{\mathfrak{B}}_\varepsilon} \cdot \xi,$$
where $\nu^{\widetilde{\mathfrak{B}}_\varepsilon}$ is the external normal of the surface $\widetilde{\mathfrak{B}}_\varepsilon$. Obviously, up to the sign, $\nu^{\widetilde{\mathfrak{B}}_\varepsilon}$ agrees with $\nu^u$. Sending $\varepsilon$ to zero, we thus get

(3.31) $$\mathcal{L}^N(\mathfrak{B}) = \mathcal{L}^N\left(\pi_\xi(\widetilde{\mathfrak{B}})\right) \geq \mathcal{L}^N\left(\pi_\xi(\widetilde{\mathfrak{B}}')\right) = \int_{\widetilde{\mathfrak{B}}'} \nu^{\widetilde{\mathfrak{B}}_\varepsilon} \cdot \xi.$$

Also, by Lemma 3.8, the exterior normal $\nu^{\widetilde{\mathfrak{B}}_\varepsilon}$ in $\widetilde{\mathfrak{B}}$ has the signed assigned by the property that $\nu^{\widetilde{\mathfrak{B}}_\varepsilon} \cdot \xi \geq 0$. Therefore,
$$|\nu^u \cdot \xi| = |\nu^{\widetilde{\mathfrak{B}}_\varepsilon} \cdot \xi| = \nu^{\widetilde{\mathfrak{B}}_\varepsilon} \cdot \xi$$
in $\widetilde{\mathfrak{B}}$. Thus, by (3.31), we get that

(3.32) $$\mathcal{L}^N(\mathfrak{B}) \geq \int_{\widetilde{\mathfrak{B}}'} |\nu^u \cdot \xi|.$$

Also, by construction, $\mathcal{Y}$ sends $\widetilde{\mathfrak{B}}$ into $\widetilde{\mathfrak{A}}$; hence, by (3.30), the change of variables formula (see, e.g., page 99 in [**18**]) and Lemma 3.6, we get that

(3.33) $$\begin{aligned} \mathcal{L}^N(\mathfrak{A}) &\leq \int_{\widetilde{\mathfrak{A}}} |\nu^{\widetilde{\mathfrak{A}}}(Y) \cdot \xi| \, dY \leq \\ &\leq \int_{\widetilde{\mathfrak{B}}'} |\nu^{\widetilde{\mathfrak{A}}}(\mathcal{Y}(X)) \cdot \xi| \, |\det D\mathcal{Y}(X)| \, dX \leq \\ &\leq \text{const} \int_{\widetilde{\mathfrak{B}}'} |\nu^{\widetilde{\mathfrak{A}}}(\mathcal{Y}(X)) \cdot \xi| \, dX. \end{aligned}$$

Thence, from (3.32), (3.29) and (3.33), we have that

$$\begin{aligned}
\mathfrak{L}^N(\mathfrak{A}) &\leq \text{const} \int_{\widetilde{\mathfrak{B}}'} |\nu^{\widetilde{\mathfrak{A}}}(\mathcal{Y}(X)) \cdot \xi| \, dX \\
&\leq \text{const} \int_{\widetilde{\mathfrak{B}}'} |\nu^u(X) \cdot \xi| \, dX \\
&\leq \mathfrak{L}^N(\mathfrak{B}) \, .
\end{aligned}$$

□

CHAPTER 4

# Measure theoretic results

This section collects some measure theory lemmata, which are extensions of analogous results in [**31**]. These lemmata will be used in the sequel for estimating the measure of the projection of the set in which a suitable barrier touches a minimal solution of (1.5). For the reader's convenience, the proofs of the lemmata of this section are deferred to the Appendix.

Given two vectors $v$ and $w$, we define $\angle(v,w)$ to be the angle[1] between these vectors, i.e.,
$$\angle(v,w) := \arccos \frac{v \cdot w}{|v|\,|w|}.$$
By elementary geometric consideration, if $|v|=|w|=1$,

(4.1) $$|v-w| \leq \angle(v,w).$$

We also define, for $l > 0$,
$$L := \left\{ (x',0,x_{N+1}) \in \mathbb{R}^{N-1} \times \mathbb{R} \times \mathbb{R} \mid |x_{N+1}| \leq 1/2 \right\},$$
$$Q_l := \left\{ (x',0,x_{N+1}) \in L \mid |x'| \leq l \right\}.$$
Of course,
$$\mathcal{L}^N(Q_l) = \mathrm{const}\, l^{N-1}.$$
For $X = (x_1, \ldots, x_{N+1}) \in \mathbb{R}^{N+1}$, and $1 \leq i \leq N+1$, we define
$$\pi_i X := (x_1, \ldots, x_{i-1}, 0, x_{i+1}, x_{N+1}).$$
Also, given $X \in \mathbb{R}^{N+1}$, we will often write $X = (x', x_N, x_{N+1}) \in \mathbb{R}^{N+1} \times \mathbb{R} \times \mathbb{R}$, i.e., the notation $x'$ will often denote the first $(N-1)$ entries of $X$.

Then, with the above notation, we have the following results:

LEMMA 4.1. *Let $u$ be a solution of (1.5). Suppose that $\mathbb{S}(Y,R)$ touches the graph of $u$ by above at $X_0 = (x_0, u(x_0)) = (x_0, g_{\mathbb{S}(Y,R)}(x_0))$. Assume that*

(4.2) $$\angle\left( \frac{\nabla u(x_0)}{|\nabla u(x_0)|}, e_N \right) \leq \frac{\pi}{8}.$$

*Then, there exists a universal $a_0 > 1$ so that, for any $a \geq a_0$, there exist a universal $\kappa > 1$ and a suitable $C > 1$, which depends only on $a$ and on universal constants, such that the following holds. For any point $Z \in L \cap B_a(\pi_N X_0)$ there exists $x$ satisfying the following properties:*

- $|x - x_0| \leq \kappa a$,
- $Z = \pi_N(x, u(x))$,
- $(x - x_0) \cdot \dfrac{\nabla u(x_0)}{|\nabla u(x_0)|} \leq H_0(u(x)) - H_0(u(x_0)) + \dfrac{C}{R},$

---

[1] As usual, the angle of the arccos ranges between 0 and $\pi$.

provided that $R$ is large enough (possibly in dependence of $a$).

LEMMA 4.2. *Let $u$ be a $C^1$-subsolution of (1.5) in $\{|x'| < l\} \times \{|x_N| < l\}$. Assume that $\mathbb{S}(Y_0, R)$ is above the graph of $u$ and that $\mathbb{S}(Y_0, R)$ touches the graph of $u$ at the point $(x_0, u(x_0))$. Suppose that*

- $|u(x_0)| < 1/2$, $|x_{0N}| < l/4$, $q := |x_0'| < l/4$;
- $\angle\left(\dfrac{\nabla u(x_0)}{|\nabla u(x_0)|}, e_N\right) \leq \dfrac{\pi}{8}.$

*Then, there exist universal constants $C_1$, $C_2 > 1 > c > 0$ such that, if*

$$q \geq C_1 \quad \text{and} \quad 4\sqrt[3]{R} \leq l \leq cR,$$

*the following holds. Let $\Xi$ be the set of points $(\mathfrak{x}, u(\mathfrak{x})) \in \mathbb{R}^N \times \mathbb{R}$ satisfying the following properties:*

- $|\mathfrak{x}'| < q/15$, $|u(\mathfrak{x})| < 1/2$, $|\mathfrak{x} - x_0| < 2q$;
- *there exists $Y \in \mathbb{R}^N \times [-1/4, 1/4]$ such that $\mathbb{S}(Y, R/C_2)$ is above $u$ and it touches $u$ at $(\mathfrak{x}, u(\mathfrak{x}))$;*
- $\angle\left(\dfrac{\nabla u(\mathfrak{x})}{|\nabla u(\mathfrak{x})|}, \dfrac{\nabla u(x_0)}{|\nabla u(x_0)|}\right) \leq \dfrac{C_1 q}{R};$
- $(\mathfrak{x} - x_0) \cdot \dfrac{\nabla u(x_0)}{|\nabla u(x_0)|} \leq \dfrac{C_1 q^2}{R} + H_0(u(\mathfrak{x})) - H_0(u(x_0)).$

*Then,*

$$\mathfrak{L}^N\left(\pi_N(\Xi)\right) \geq cq^{N-1}.$$

LEMMA 4.3. *Let $C \geq 2$, $c > 0$. Let us consider, for $k \in \mathbb{N}$, a family of sets $D_k \subseteq L$, so that $D_k \subseteq D_{k+1}$ for any $k \in \mathbb{N}$. Assume that the following properties hold, for some $l > Ca \geq 2a > c > 0$:*

(P1) $D_0 \cap Q_l \neq \emptyset$;

(P2) *for any $Z_0 \in D_k \cap Q_{2l}$ and any $Z_1 \in L$, with $a \leq |Z_1 - Z_0| \leq 2l$, one has that*

$$\mathfrak{L}^N\left(D_{k+1} \cap B_{|Z_1-Z_0|/10}(Z_1)\right) \geq c\,\mathfrak{L}^N\left(L \cap B_{|Z_1-Z_0|}(Z_1)\right).$$

*Define, for any $k \in \mathbb{N}$*

(4.3) $$E_k := \left\{Z \in L \mid \text{dist}(Z, D_k) \leq a\right\}.$$

*Then, there exists $a_0 > 1 > c_0 > 0$ universal constants and $c^\star > 0$, which depends only on $a$, $c$ and universal constants, such that*

$$\mathfrak{L}^N\left(Q_l \setminus E_k\right) \leq (1 - c^\star)\,\mathfrak{L}^N(Q_l),$$

*provided that $a \geq a_0$ and $c$, $C^{-1} \in (0, c_0]$.*

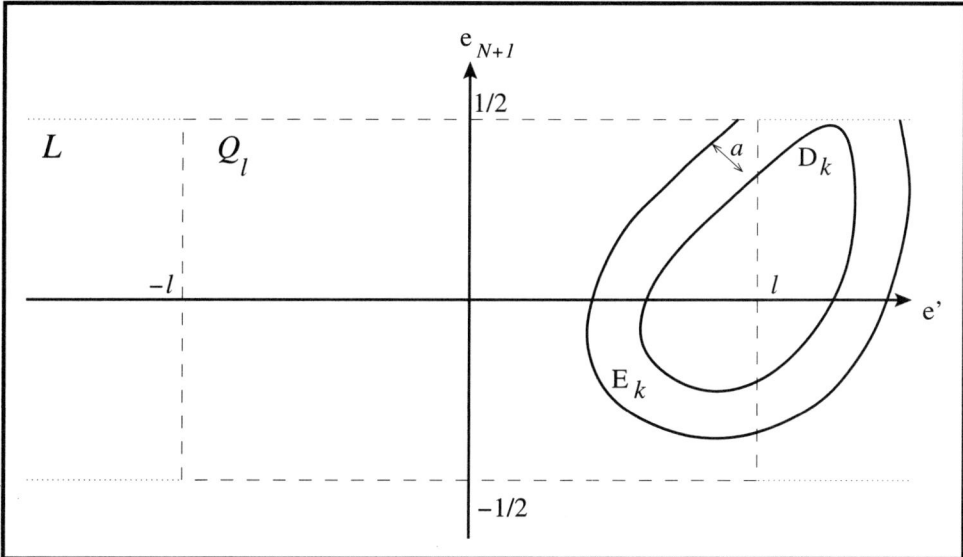

**The covering sets of Lemma 4.3**

Full details of the proofs of the above lemmata will be provided in the Appendix.

# CHAPTER 5

# Estimates on the measure of the projection of the contact set

We now show how to use Proposition 3.14 and the measure theoretic lemmata stated in §4 in order to deduce a measure estimate on the projection of the contact sets between barriers and minimal solutions of (1.5). To this aim, we first need an estimate on the contact sets obtained by touching $u$ by above "for the first time", as dealt with by the following result.

LEMMA 5.1. *Let $C, C' > 1$ be suitably large constants. Let $K_l := \{|x'| < Cl\} \times \{|x_N| < Cl\}$. Let $u \in W^{1,p}(K_l)$ be a local minimizer for $\mathcal{F}$ in $K_l$. Assume that $u(0) = 0$ and that $u(x) < 0$ if $x_N < -\theta$, for some $\theta > 0$. Define $R_0 := l^2/(C\theta)$. Let $\Xi$ be the set of points $(\mathfrak{x}, u(\mathfrak{x})) \in K_l$ satisfying:*

- $|\mathfrak{x}'| \leq l$, $|u(\mathfrak{x})| < 1/2$;
- *there exists $Y \in \mathbb{R}^N \times [-1/4, 1/4]$ such that $\mathbb{S}(Y, R_0)$ is above the graph of $u$ in $\{|x'| < C'l\} \times \{|x_N| < C'l\}$ and it touches the graph of $u$ at $(\mathfrak{x}, u(\mathfrak{x}))$;*
- $\angle\left(\dfrac{\nabla u(\mathfrak{x})}{|\nabla u(\mathfrak{x})|}, e_N\right) \leq \dfrac{8l}{R_0}$;
- $\mathfrak{x}_N \leq \dfrac{\theta}{4} + H_0(u(\mathfrak{x}))$.

*Then, there exists a universal constant $c > 0$ such that, for any $\theta_0 > 0$ there exists $\varepsilon_0(\theta_0) > 0$ for which, if*

$$\frac{\theta}{l} \leq \varepsilon_0(\theta_0), \qquad \theta \geq \theta_0,$$

*one has that*

$$\mathcal{L}^N\left(\pi_N(\Xi)\right) \geq c\, l^{N-1}.$$

PROOF. Exploiting Lemma 2.23, we have that, if $C$ is large enough,

(5.1) $$u(x) \leq g_l(x_N + \theta)$$

for any $x$ so that $|x'| \leq C'l$ and $|x_N| \leq C'l$, with $C'$ large if so is $C$. Let us define

(5.2) $$R_0 := l^2/(C\theta)$$

and, for $C'' > 0$ conveniently large, let us consider the set

$$\widetilde{\mathfrak{H}} := \Big\{ Y = (y, y_{N+1}) \in \mathbb{R}^{N+1} \text{ such that }$$
$$|y'| \leq l/C'', \ |y_{N+1}| \leq 1/4$$
$$\text{and so that, if } (0,\ldots,0,x_N,0) \in \mathbb{S}(Y, R_0) \text{ then } x_N \leq 0 \Big\}.$$

We claim that

(5.3) $$g_{\mathbb{S}(Y,R_0)}(x) > g_l(x_N + \theta) \text{ for any } Y \in \widetilde{\mathfrak{D}},$$
provided that $x \in K_l$ and $|x'| \in (l, C'l)$.

To prove this, let $Y \in \widetilde{\mathfrak{D}}$ and define
$$\Sigma := \{g_{\mathbb{S}(Y,R_0)} = 0\} = \mathbb{S}(Y, R_0) \cap \{x_{N+1} = 0\}.$$

Then, the last condition in the definition of $\widetilde{\mathfrak{D}}$ reads

(5.4) $$\text{if } (0, \ldots, 0, x_N) \in \Sigma, \text{ then } x_N \leq 0.$$

Recalling the definitions on page 14 and (2.39), one sees that $\Sigma$ is an $(N-1)$-dimensional sphere, namely
$$\Sigma = \{|x - y| = r\}$$
with

(5.5) $$r := R_0 - H_0(y_{N+1}) - \frac{\overline{C}_0}{2R_0} y_{N+1}^2.$$

Let us now estimate $r$ by noticing that, if $l$ (and therefore $R_0$) is suitably large, we have that

(5.6) $$\frac{2l^2}{3C\theta} = \frac{2}{3}R_0 \leq R_0 - \text{const} \leq$$
$$\leq r \leq$$
$$\leq R_0 + \text{const} \leq \frac{3}{2}R_0 = \frac{3l^2}{2C\theta}.$$

Notice also that $x$ in (5.3) must lie in the intersection between $K_l$ and the domain of $g_{\mathbb{S}(Y,R_0)}$, otherwise there is nothing to prove; therefore,
$$|x - y| \leq \text{const}\,(C'l + R_0) \leq \text{const}\,R_0,$$
and so, by (5.6),

(5.7) $$|x - y| \leq \text{const}\,r.$$

We now point out that $\Sigma$ is below the hyperplane $x_N = \theta/8$, that is

(5.8) $$x_N \leq \theta/8, \text{ for any } x \in \Sigma.$$

In order to prove (5.8), let
$$\bar{y} := y - y_1 e_1 - \cdots - y_{N-1} e_{N-1} = (0, \ldots, 0, y_N),$$
so that, by the definition of $\widetilde{\mathfrak{D}}$,

(5.9) $$|y - \bar{y}| = |y'| \leq \frac{l}{C''},$$

which is less than $r$ due to (5.6), provided that $\theta/l$ is small enough. Thus, let $\tilde{t} > 0$ be so that
$$\tilde{y} := \bar{y} + \tilde{t} e_N \in \Sigma.$$

## 5. ESTIMATES ON THE MEASURE OF THE PROJECTION OF THE CONTACT SET

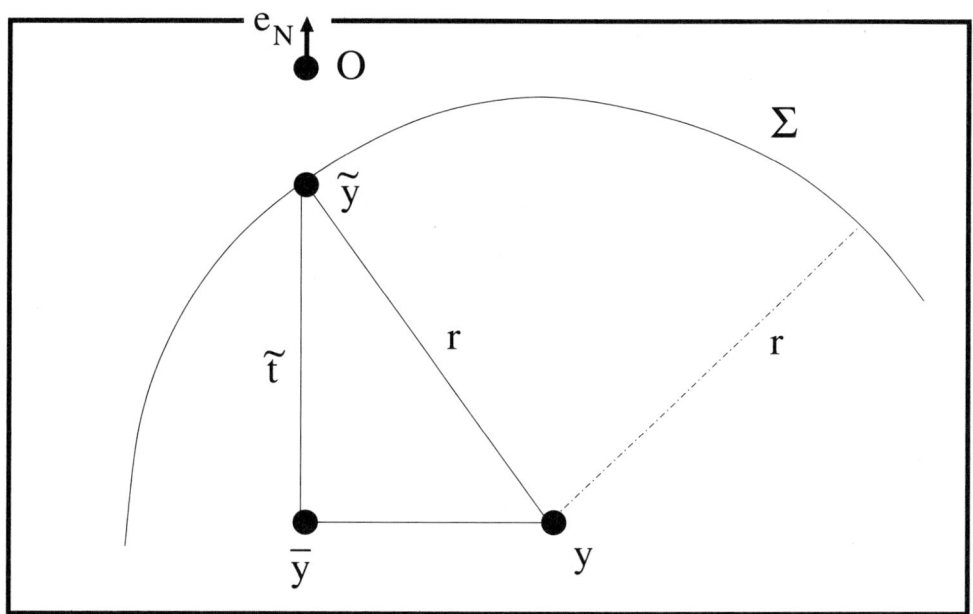

**The geometry related with $\tilde{y}$**

Then $\tilde{y} = (0, \ldots, 0, y_N + \tilde{t})$, thus, from (5.4),

(5.10)   $$y_N \leq -\tilde{t}.$$

Also, from (5.9),

$$\tilde{t}^2 = r^2 - |y - \bar{y}|^2 \geq r^2 - \left(\frac{l}{C''}\right)^2,$$

therefore, in the light of (5.10) and (5.6),

(5.11)
$$\begin{aligned} y_N + r &\leq r - \tilde{t} \leq \\ &\leq r - \sqrt{r^2 - \left(\frac{l}{C''}\right)^2} \leq \\ &\leq \frac{\theta}{8}, \end{aligned}$$

provided that $C''$ is large enough, completing the proof of (5.8).

Let us now go back to the proof of (5.3). For this, we introduce the following notation: define

$$\begin{aligned} d_1(x) &:= |x - y| - r, \\ d_2(x) &:= x_N + \theta. \end{aligned}$$

Let now $x$ be as requested in (5.3). From (5.7),

(5.12)   $$0 \leq r + d_1(x) \leq \operatorname{const} r.$$

Also,

$$|x' - y'| \geq |x'| - |y'| \geq \frac{l}{2} - \frac{l}{C'} \geq \frac{2}{5}l,$$

if $C' \geq 10$, thus,

$$\left(r + d_1(x)\right)^2 = |x-y|^2 =$$
$$= |x'-y'|^2 + |x_N - y_N|^2 \geq$$
(5.13)
$$\geq \frac{4}{25}l^2 + |x_N - y_N|^2;$$

thus, from (5.13) and (5.11), we infer that

$$x_N \leq y_N + \sqrt{\left(r+d_1(x)\right)^2 - \frac{4}{25}l^2} \leq$$
$$\leq \frac{\theta}{8} - r + \sqrt{\left(r+d_1(x)\right)^2 - \frac{4}{25}l^2}.$$

This, (5.12) and (5.6) imply that

$$x_N \leq -2\theta + d_1(x)$$

and, therefore,

$$d_1(x) \geq x_N + 2\theta = d_2(x) + \theta,$$

proving that

(5.14) $$d_1 \geq d_2 + \theta$$

in $K_l \cap \{|x'| \in (l, C'l)\}$.

We now observe that

(5.15) $$H_{y_{N+1}, R_0}(s) - H_{y_{N+1}, R_0}(0) \leq H_0(s) + \frac{2\overline{C}_0}{R_0},$$

for any $s \in [s_{R_0}, 1]$. To prove this, recall Definition 2.8 to get

$$H_{y_{N+1}, R_0}(s) - H_{y_{N+1}, R_0}(0) = \int_0^s \frac{(p-1)^{\frac{1}{p}}}{(p\, h_{s_0, R}(\zeta))^{\frac{1}{p}}} d\zeta,$$

and use Definition 2.5 and (2.2) to deduce (5.15).

Therefore, from (5.15) and (5.2), if $l$ is large enough, we get that

(5.16) $$H_{y_{N+1}, R_0}(s) - H_{y_{N+1}, R_0}(0) < H_0(s) + \frac{\theta}{2}.$$

Notice now that, by (2.7), (5.2) and the definition of $s_l$ given in Lemma 2.20, we have that

$$s_{R_0} \geq -1 + \frac{\text{const}}{R_0^{1/p}} = -1 + \frac{\text{const}\, \theta^{1/p}}{l^{2/p}} > -1 + s_l.$$

In particular, the function

$$h_{y_{N+1}, R_0}(s) - h_l(s)$$

is defined for any $s$ so that

$$s_{R_0} = \max\{s_{R_0}, -1+s_l\} \leq s \leq 1.$$

Also, if $c_* > 0$ is suitably small (possibly in dependence also of $\theta_0$) and

$$s_{R_0} \leq s \leq -1 + c_*/l^{1/p},$$

## 5. ESTIMATES ON THE MEASURE OF THE PROJECTION OF THE CONTACT SET

we infer from (2.6), the definition of $h_l$ given in Lemma 2.20 and the one of $h_{s_0,R}$ given on page 10 that

$$h_{y_{N+1},R_0}(s) - h_l(s) =$$
$$= -h_0(s_R) - \frac{\widehat{C}_0}{R_0}(s - s_{R_0}) +$$
$$+ h_0(s_l - 1) + \frac{\overline{C}_2}{l}\left((1+s)^p - s_l^p\right) \leq$$
$$\leq -\frac{1}{R_0} + e^{-\operatorname{const} l} + \frac{\overline{C}_2 c_*^p}{l^2} \leq$$
$$\leq -\frac{C\theta_0}{l^2} + e^{-\operatorname{const} l} + \frac{\overline{C}_2 c_*^p}{l^2} \leq 0,$$

that is

(5.17) $\qquad h_{y_{N+1},R_0}(s) \leq h_l(s)$ for any $s \in [s_{R_0}, -1 + c_*/l^{1/p}]$,

provided that $c_*$ is small enough. Analogously, one can show that

(5.18) $\qquad h_{y_{N+1},R_0}(s) \geq h_l(s)$ for any $s \in [1 - c_*/l^{1/p}, 1]$.

From (5.17), (5.18) and the definitions of $H_{y_{N+1},R_0}$ and $H_l$ (see pages 13 and 24), we deduce that the maximum of the function

$$[s_{R_0}, 1] \ni s \mapsto H_{y_{N+1},R_0}(s) - H_l(s)$$

occurs for $|s| \leq 1 - c_*/l^{1/p}$. For these values of $s$, estimate (2.61) in Lemma 2.20 implies that

$$H_0(s) \leq H_l(s) - \frac{\operatorname{const}}{l} \log(1 - |s|) \leq$$
$$\leq H_l(s) + \frac{\operatorname{const}}{l} \log \frac{l^{1/p}}{c_*} \leq$$
$$\leq H_l(s) + \frac{\theta_0}{2},$$

provided that $l$ is suitably large. Thus, summarizing the above observations and using (5.16), we have that

$$\max_{[s_{R_0}, 1]} \left(H_{y_{N+1},R_0} - H_l\right) =$$
$$= \max_{[-1+c_*/l^{1/p}, 1-c_*/l^{1/p}]} \left(H_{y_{N+1},R_0} - H_l\right) <$$
$$< \max_{[-1+c_*/l^{1/p}, 1-c_*/l^{1/p}]} \left(H_0 - H_l\right) + \frac{\theta}{2} + H_{y_{N+1},R_0}(0) \leq$$
$$\leq \frac{\theta_0}{2} + \frac{\theta}{2} + H_{y_{N+1},R_0}(0).$$

Hence,

(5.19) $\qquad H_{y_{N+1},R_0}(s) - H_{y_{N+1},R_0}(0) < H_l(s) + \theta$,

for any $s \in (s_{R_0}, 1]$. From (5.19), by inverting $H_{y_{N+1},R_0}$, we have

$$s < g_{y_{N+1},R_0}\left(H_{y_{N+1},R_0}(0) + H_l(s) + \theta\right),$$

for any $s \in (s_{R_0}, 1]$ and so, for $s := g_l(d_1(x) - \theta)$, we get
$$g_l(d_1(x) - \theta) < g_{y_{N+1}, R_0}\Big(d_1(x) + H_{y_{N+1}, R_0}(0)\Big).$$
Therefore, recalling also (5.14),
$$g_{y_{N+1}, R_0}\Big(|x - y| - r + H_{y_{N+1}, R_0}(0)\Big) =$$
$$= g_{y_{N+1}, R_0}\Big(d_1(x) + H_{y_{N+1}, R_0}(0)\Big) >$$
$$> g_l(d_1(x) - \theta) \geq$$
$$\geq g_l(d_2(x)) =$$
$$= g_l(x_N + \theta).$$
This completes the proof of (5.3).

By (5.3) and (5.1), we have that
(5.20)
$$g_{\mathbb{S}(Y, R_0)}(x) > u(x) \text{ for any } Y \in \widetilde{\mathfrak{D}}, \text{ provided that } |x'| \in (l, C'l) \text{ and } |x_N| \leq C'l.$$

Let now $\widetilde{\Xi}$ be the set of $(\mathfrak{x}, u(\mathfrak{x}))$'s described in the statement of Lemma 5.1. Let us also define $\Xi := \pi_{e_N}\widetilde{\Xi}$ and $\mathfrak{D} := \pi_{e_N}\widetilde{\mathfrak{D}}$. Of course,
$$\mathfrak{D} = \Big\{Y = (y', 0, y_{N+1}) \in \mathbb{R}^{N+1} \text{ such that } |y'| \leq l/C'', \ |y_{N+1}| \leq 1/4\Big\},$$
therefore
(5.21)
$$\mathfrak{L}^N(\mathfrak{D}) \geq \text{const } l^{N-1}.$$

For any $Y \in \mathfrak{D}$, from (5.1) and the fact that $\mathbb{S}(Y, R_0)$ takes value 1 on the boundary of its domain of definition, we know that $\mathbb{S}(Y - te_N, R_0)$ is above the graph of $u$ in the intersection between $\{|x'| \leq C'l\} \times \{|x_N| \leq C'l\}$ and the domain of definition of $\mathbb{S}(Y - te_N, R_0)$, provided that $t$ is large enough. Also, by looking at the construction of $\bar{y}$ on page 48, it follows easily, by decreasing $t$, that there will be a suitable $t^*$ for which $\mathbb{S}(Y - t^*e_N, R_0)$ touches for the first time the graph of $u$, say at the point $\widetilde{X}$. We denote by $\widetilde{\mathfrak{G}}$ the set of such touching points $\widetilde{X}$'s and define also $\mathfrak{G} := \pi_{e_N}\widetilde{\mathfrak{G}}$.

We claim that
(5.22)
$$\mathfrak{G} \subseteq \Xi.$$
For proving this, take any $\widetilde{X} \in \widetilde{\mathfrak{G}}$ be a touching point between $\mathbb{S}(Y - t^*e_N, R_0)$ and the graph of $u$, as described above. Let us observe that, since $u(0) = 0$, the first touching property of $\widetilde{X}$ implies that if $\check{X} = (0, \ldots, \check{x}_N, 0) \in \mathbb{S}(Y - t^*e_N, R_0)$, then $\check{x}_N \leq 0$, hence
$$\widetilde{\mathfrak{D}} \ni Y - t^*e_N =: \widetilde{Y}.$$
From this and (5.20), we gather that
(5.23)
$$|\widetilde{x}'| \leq l.$$
We now show that
(5.24)
$$\widetilde{x} \text{ is in the interior of } \{|x'| \leq C'l\} \times \{|x_N| \leq C'l\}.$$
Note that, thanks to (5.23), this will be proved if we show that $|\widetilde{x}_N| < C'l$.

## 5. ESTIMATES ON THE MEASURE OF THE PROJECTION OF THE CONTACT SET

Let us first show that $\widetilde{x}_N > -C'l$. If, by contradiction, $\widetilde{x}_N = -C'l$, we gather from (2.7) and (5.1) that

$$
\begin{aligned}
-1 + \frac{\mathrm{const}\, \theta_0^{1/p}}{l^{2/p}} &\leq -1 + \frac{\mathrm{const}}{R_0^{1/p}} \leq \\
&\leq s_{R_0} \leq \\
&\leq g_{\mathbb{S}(\widetilde{Y},R_0)}(\widetilde{x}) = \\
&= u(\widetilde{x}) \leq \\
&\leq g_l(\widetilde{x}_N + \theta) \leq \\
&\leq g_l\left(-\frac{C'l}{2}\right) = \\
&= -1 + e^{-\mathrm{const}\, l},
\end{aligned}
$$

which is a contradiction for large $l$. This shows that $\widetilde{x}_N > -C'l$ and thus we now show that $\widetilde{x}_N < C'l$, in order to complete the proof of (5.24). That $\widetilde{x}_N < C'l$ will be actually obtained from the fact that the domain of $\mathbb{S}(\widetilde{Y}, R_0)$ is below the hyperplane $\{x_N \leq l/2\}$. To prove this, first note that, by (5.11), we have that

$$(5.25) \qquad \widetilde{y}_N \leq -r + \frac{\theta}{8},$$

Also, if $\mathfrak{x}$ is in the domain of $\mathbb{S}(\widetilde{Y}, R_0)$, we have that

$$\mathfrak{x}_N \leq \widetilde{y}_N + H_{\widetilde{y}_{N+1},R_0}(1) - H_0(\widetilde{y}_{N+1}) + R_0.$$

Thus, (5.25), (5.5) and (2.16) yield that

$$\mathfrak{x}_N \leq \mathrm{const}\,(1 + \log R_0) \leq \frac{l}{2},$$

hence the domain of $\mathbb{S}(\widetilde{Y}, R_0)$ is below $\{x_N \leq l/2\}$ and therefore $\widetilde{x}_N < C'l$.

This ends the proof of (5.24).

Proposition 2.13 and (5.24) yield that

$$(5.26) \qquad |u(\widetilde{x})| \leq 1/2.$$

We now notice that, from (5.26) and (2.20),

$$
\begin{aligned}
-\frac{1}{2} &\leq u(\widetilde{x}) = \\
&= g_{\mathbb{S}(\widetilde{Y},R_0)}(\widetilde{x}) = \\
&= g_{\widetilde{y}_{N+1},R_0}\left(H_0(\widetilde{y}_{N+1}) + |\widetilde{x} - \widetilde{y}| - R_0\right),
\end{aligned}
$$

and so, by Definition 2.11,

$$H_0(\widetilde{y}_{N+1}) + |\widetilde{x} - \widetilde{y}| - R_0 \geq H_{\widetilde{y}_{N+1},R_0}(-1/2),$$

from which we deduce that

$$(5.27) \qquad |\widetilde{x} - \widetilde{y}| \geq R_0 - \mathrm{const} \geq R_0/2,$$

54   5. ESTIMATES ON THE MEASURE OF THE PROJECTION OF THE CONTACT SET

provided that $l$ (and so $R_0$) is large enough. On the other hand, exploiting (5.23) and the definition of $\widetilde{\mathfrak{D}}$ given on page 47, we have that

$$|\widetilde{x}' - \widetilde{y}'| \leq |\widetilde{x}'| + |\widetilde{y}'| \leq l + \frac{l}{C''} \leq 2l.$$

Hence, from (5.27),

$$\frac{|\widetilde{x}_N - \widetilde{y}_N|^2}{|\widetilde{x} - \widetilde{y}|^2} = 1 - \frac{|\widetilde{x}' - \widetilde{y}'|^2}{|\widetilde{x} - \widetilde{y}|^2} \geq$$
$$\geq 1 - \frac{16l^2}{R_0^2},$$

and, therefore,

$$1 - \frac{1}{4}\left[\angle\left(\frac{\widetilde{x} - \widetilde{y}}{|\widetilde{x} - \widetilde{y}|}, e_N\right)\right]^2 \geq \cos^2\left[\angle\left(\frac{\widetilde{x} - \widetilde{y}}{|\widetilde{x} - \widetilde{y}|}, e_N\right)\right] =$$
$$= \frac{|\widetilde{x}_N - \widetilde{y}_N|^2}{|\widetilde{x} - \widetilde{y}|^2} \geq$$
$$\geq 1 - \frac{16l^2}{R_0^2},$$

that is

(5.28) $$\angle\left(\frac{\widetilde{x} - \widetilde{y}}{|\widetilde{x} - \widetilde{y}|}, e_N\right) \leq \frac{8l}{R_0}.$$

Moreover, from the touching property of $\widetilde{X}$ and (2.20), we have that

$$\angle\left(\frac{\nabla u(\widetilde{x})}{|\nabla u(\widetilde{x})|}, e_N\right) = \angle\left(\frac{\nabla g_{\mathbb{S}(\widetilde{Y},R_0)}(\widetilde{x})}{|\nabla g_{\mathbb{S}(\widetilde{Y},R_0)}(\widetilde{x})|}, e_N\right) =$$
$$= \angle\left(\frac{\widetilde{x} - \widetilde{y}}{|\widetilde{x} - \widetilde{y}|}, e_N\right).$$

Therefore, from (5.28),

(5.29) $$\angle\left(\frac{\nabla u(\widetilde{x})}{|\nabla u(\widetilde{x})|}, e_N\right) \leq \frac{8l}{R_0}.$$

Furthermore, recalling (2.39),

$$H_{\widetilde{y}_{N+1},R_0}(u(\widetilde{x})) - H_{\widetilde{y}_{N+1},R_0}(0) =$$
$$= H_{\widetilde{y}_{N+1},R_0}\left(g_{\mathbb{S}(\widetilde{Y},R_0)}(\widetilde{x})\right) - H_{\widetilde{y}_{N+1},R_0}(0) =$$
$$= H_{\widetilde{y}_{N+1},R_0}\left(g_{\mathbb{S}(\widetilde{Y},R_0)}(\widetilde{x})\right) + \frac{\overline{C_0}}{2R_0}\widetilde{y}_{N+1}^2,$$

# 5. ESTIMATES ON THE MEASURE OF THE PROJECTION OF THE CONTACT SET

thus, from (2.33) and (2.35),

$$\begin{aligned}
H_{\widetilde{y}_{N+1}, R_0}(u(\widetilde{x})) &- H_{\widetilde{y}_{N+1}, R_0}(0) = \\
= H_{\widetilde{y}_{N+1}, R_0} &\Big( \rho_{\widetilde{y}_{N+1}, R_0}(H_0(\widetilde{y}_{N+1}) + |\widetilde{x} - \widetilde{y}| - R_0) \Big) + \\
+ \frac{\overline{C}_0}{2R_0} &\widetilde{y}_{N+1}^2 = \\
= H_0(\widetilde{y}_{N+1}) &+ |\widetilde{x} - \widetilde{y}| - R_0 + \frac{\overline{C}_0}{2R_0} \widetilde{y}_{N+1}^2 .
\end{aligned}$$

Hence, by (5.5),

$$(5.30) \qquad \begin{aligned} H_{\widetilde{y}_{N+1}, R_0}(u(\widetilde{x})) - H_{\widetilde{y}_{N+1}, R_0}(0) &= \\ = |\widetilde{x} - \widetilde{y}| - r\,. \end{aligned}$$

We now claim that

$$(5.31) \qquad \widetilde{x}_N \leq H_0(u(\widetilde{x})) + \frac{\theta}{4}.$$

For proving this, we denote by $\widehat{x}$ the intersection point between the sphere $\{g_{\mathbb{S}(\widetilde{Y}, R_0)} = 0\}$ and the half-line from $\widetilde{y}$ towards $\widetilde{x}$. Then, by (5.8),

$$\widehat{x}_N \leq \theta/8\,.$$

We now distinguish two cases: either $\widehat{x}$ is inside or it is outside the sphere $\{g_{\mathbb{S}(\widetilde{Y}, R_0)} = 0\}$. If it is inside, then

$$\begin{aligned}
|\widetilde{x} - \widetilde{y}| - r &= |\widehat{x} - \widetilde{x}| \geq \\
&\geq |\widehat{x}_N - \widetilde{x}_N| \geq \\
&\geq \widetilde{x}_N - \widehat{x}_N \geq \\
&\geq \widetilde{x}_N - \theta/8\,.
\end{aligned}$$

Thus, from the latter estimate, (5.30) and (5.15), we have that

$$\begin{aligned}
\widetilde{x}_N &\leq |\widetilde{x} - \widetilde{y}| - r + \frac{\theta}{8} = \\
&= H_{\widetilde{y}_{N+1}}(u(\widetilde{x})) - H_{\widetilde{y}_{N+1}}(0) + \frac{\theta}{8} \leq \\
&\leq H_0(u(\widetilde{x})) + \frac{2\overline{C}_0}{R_0} + \frac{\theta}{8}\,.
\end{aligned}$$

Therefore, if $l$ (and so $R_0$) is large enough, (5.31) follows in this case. Let us now deal with the case in which $\widehat{x}$ is outside the sphere $\{g_{\mathbb{S}(\widetilde{Y}, R_0)} = 0\}$. By (5.29), we infer in this case that $\widehat{x}_N \geq \widetilde{x}_N$ and that

$$\begin{aligned}
\widehat{x}_N - \widetilde{x}_N &= |\widehat{x}_N - \widetilde{x}_N| = \\
&= |\widehat{x} - \widetilde{x}| \cos\Big(\angle(\widehat{x} - \widetilde{x}, e_N)\Big) \geq \\
&\geq |\widehat{x} - \widetilde{x}| \left(1 - \frac{\operatorname{const} l^2}{R_0^2}\right).
\end{aligned}$$

56   5. ESTIMATES ON THE MEASURE OF THE PROJECTION OF THE CONTACT SET

Therefore, (5.30) and (5.15) yield that

$$\begin{aligned}
H_0(u(\widetilde{x})) + \frac{2\overline{C}_0}{R_0} &\geq |\widetilde{y} - \widetilde{x}| - r = \\
&= -|\widehat{x} - \widetilde{x}| \geq \\
&\geq \frac{\widetilde{x}_N - \widehat{x}_N}{1 - \frac{\text{const } l^2}{R_0^2}} \geq \\
&\geq \frac{\widetilde{x}_N - (\theta/8)}{1 - \frac{\text{const } l^2}{R_0^2}},
\end{aligned}$$

which easily implies (5.31) in this case. This completes the proof of (5.31).

Thus, in the light of (5.23), (5.26), (5.29) and (5.31), we have that $\widetilde{X} \in \widetilde{\Xi}$ and, therefore, that $\pi_{e_N} \widetilde{X} \in \Xi$, ending the proof of (5.22).

Now we exploit Proposition 3.14, applied to $\mathfrak{G}$ and $\mathfrak{D}$: from that, (5.22) and (5.21),

$$\begin{aligned}
\mathfrak{L}^N(\Xi) &\geq \mathfrak{L}^N(\mathfrak{G}) \geq \\
&\geq \text{const } \mathfrak{L}^N(\mathfrak{D}) \geq \\
&\geq \text{const } l^{N-1}.
\end{aligned}$$

This completes the proof of Lemma 5.1.    □

The next one is the main result of this section:

PROPOSITION 5.2. *Let $C$ be a suitably large constant. Let $K_l := \{|x'| < C\, l\} \times \{|x_N| < C\, l\}$. Let $u \in W^{1,p}(K_l)$ be a local minimizer for $\mathcal{F}$ in $K_l$. Assume that $u(0) = 0$ and that $u(x) < 0$ if $x_N < -\theta$, for some $\theta > 0$. Fix $\bar{C} > 0$ and $k \in \mathbb{N}$. Let $\Xi$ be the set of points $(\mathfrak{x}, u(\mathfrak{x}))$ satisfying the following properties:*

- $|\mathfrak{x}'| \leq l,\ |\mathfrak{x}_{N+1}| \leq 1/2$;
- $\mathfrak{x}_N \leq \bar{C}^k \theta + H_0(u(\mathfrak{x}))$.

*Then, there exist positive universal constants $c$ and $\hat{c}$ for which the following holds. For any $\theta_0 > 0$, there exists $\varepsilon_0(\theta_0) > 0$, so that, if*

$$\frac{\theta}{l} \leq \varepsilon_0(\theta_0), \qquad \theta \geq \theta_0 \quad \text{and} \quad \frac{\bar{C}^k \theta}{l} \leq \hat{c},$$

*then*

$$\mathfrak{L}^N\big(\pi_N(\Xi)\big) \geq \big(1 - (1-c)^k\big)\, \mathfrak{L}^N(Q_l).$$

PROOF. Let $R_0 := l^2/(C\theta)$, with $C$ suitably large. For any $k \in \mathbb{N}$, let $R_k := R_0\, \bar{C}^{-k}$, where $\bar{C}$ is a positive universal constant, to be chosen suitably large in the sequel. We define $\mathfrak{D}_k \subseteq \mathbb{R}^{N+1}$ as the set of points $(\mathfrak{x}, u(\mathfrak{x}))$ satisfying the following properties:

- $|\mathfrak{x}'| \leq C\, l/2,\ |u(\mathfrak{x})| < 1/2$;
- there exists $Y \in \mathbb{R}^N \times [-1/4, 1/4]$ so that $\mathbb{S}(Y, R_k)$ is above the graph of $u$ in $\{|x'| < C\, l/2\} \times \{|x_N| < C\, l/2\}$ and it touches the graph of $u$ at $(\mathfrak{x}, u(\mathfrak{x}))$;
- $\angle\left(\dfrac{\nabla u(\mathfrak{x})}{|\nabla u(\mathfrak{x})|}, e_N\right) \leq \dfrac{\bar{C}^k l}{R_0}$;

## 5. ESTIMATES ON THE MEASURE OF THE PROJECTION OF THE CONTACT SET

- $\mathfrak{x}_N \leq \dfrac{\bar{C}^k \theta}{4} + H_0(u(\mathfrak{x}))$.

We also set $D_k := \pi_N(\mathfrak{D}_k)$. We would like to apply Lemma 4.3 to $D_k$, and we therefore now prove that $D_k$ fulfills the assumption of Lemma 4.3. For this, first of all, notice that, by Lemma 5.1,

$$\text{(5.32)} \qquad D_0 \cap Q_l \neq \emptyset.$$

Let us now fix $Z_k \in D_k \cap Q_{2l}$. By construction, there exists $(x_k, u(x_k)) \in \mathfrak{D}_k$ so that $Z_k = \pi_N(x_k, u(x_k))$. Take also $\tilde{Z} \in L$, with $a \leq |\tilde{Z} - Z_k| =: q \leq 2l$, and suppose $a$ suitably large. We claim that

$$\text{(5.33)} \qquad \mathfrak{L}^N\left(D_{k+1} \cap B_{q/10}(\tilde{Z})\right) \geq \mathfrak{L}^N\left(L \cap B_q(\tilde{Z})\right).$$

In order to prove the above inequality, we denote by $\hat{C} > 0$ a constant, to be suitably chosen in the sequel, and we define $\tilde{\Xi}$ as the set of points $(\mathfrak{x}, u(\mathfrak{x}))$ satisfying the following properties:

- $|\mathfrak{x}' - \tilde{z}'| \leq q/15$, $|\mathfrak{x} - x_k| < 4l$, $|u(\mathfrak{x})| < 1/2$;
- there exists $Y \in \mathbb{R}^N \times [-1/4, 1/4]$ so that $\mathbb{S}(Y, R_{k+1})$ is above the graph of $u$ in $\{|x' - \tilde{z}'| < C\, l/2\} \times \{|x_N| < C\, l/2\}$ and it touches the graph of $u$ at $(\mathfrak{x}, u(\mathfrak{x}))$;
- $\angle\left(\dfrac{\nabla u(\mathfrak{x})}{|\nabla u(\mathfrak{x})|}, \dfrac{\nabla u(x_k)}{|\nabla u(x_k)|}\right) \leq \dfrac{\hat{C} \bar{C}^k l}{R_0}$;
- $(\mathfrak{x} - x_k) \cdot \dfrac{\nabla u(x_k)}{|\nabla u(x_k)|} \leq \dfrac{\hat{C} \bar{C}^k l^2}{4 R_0} + H_0(u(\mathfrak{x})) - H_0(u(x_k))$.

Notice that, by means[1] of Lemma 4.2 (applied in $\{|x' - \tilde{z}'| \leq 8l\} \times \{|x_N| \leq 8l\}$),

$$\text{(5.34)} \qquad \mathfrak{L}^N(\pi_N(\tilde{\Xi}) \cap B_{q/10}(\tilde{Z})) \geq \text{const } q^{N-1} \geq \text{const } \mathfrak{L}^N\left(L \cap B_q(\tilde{Z})\right).$$

Let us now deduce some easy properties of $\tilde{\Xi}$. First of all, by the definitions of $\tilde{\Xi}$ and $\mathfrak{D}_k$, we have that, for any $(\mathfrak{x}, u(\mathfrak{x})) \in \tilde{\Xi}$,

$$\angle\left(\dfrac{\nabla u(\mathfrak{x})}{|\nabla u(\mathfrak{x})|}, e_N\right) \leq \angle\left(\dfrac{\nabla u(\mathfrak{x})}{|\nabla u(\mathfrak{x})|}, \dfrac{\nabla u(x_k)}{|\nabla u(x_k)|}\right) + \angle\left(\dfrac{\nabla u(x_k)}{|\nabla u(x_k)|}, e_N\right) \leq$$

$$\leq \dfrac{\hat{C} \bar{C}^k l}{R_0} + \dfrac{\bar{C}^k l}{R_0} \leq$$

$$\text{(5.35)} \qquad \leq \dfrac{\bar{C}^{k+1} l}{R_0},$$

---

[1] We apply here Lemma 4.2 with $R_k$ replacing what there was denoted by $R$. Note also that

$$\sqrt{|\tilde{Z} - Z_k|^2 - 4} \leq |x'_k - \tilde{z}'_k| \leq \sqrt{|\tilde{Z} - Z_k|^2 + 4},$$

thus

$$|x'_k - \tilde{z}'_k| \in \left(\dfrac{9}{10} q, \dfrac{11}{10} q\right),$$

if $a$ is large enough. Finally, observe that, by construction,

$$\pi_N(\tilde{\Xi}) \subseteq B_{q/10}(\tilde{Z}).$$

## 58   5. ESTIMATES ON THE MEASURE OF THE PROJECTION OF THE CONTACT SET

provided that $\bar{C}$ is big enough with respect to $\hat{C}$. Furthermore, by (4.1) and the definition of $\mathfrak{D}_k$,

$$
\text{(5.36)} \quad \left| \frac{\nabla u(x_k)}{|\nabla u(x_k)|} - e_N \right| \leq \angle\left( \frac{\nabla u(x_k)}{|\nabla u(x_k)|}, e_N \right) \leq \frac{\bar{C}^k l}{R_0},
$$

and so

$$
\text{(5.37)} \quad \left| (\mathfrak{x} - x_k) \cdot \left( \frac{\nabla u(x_k)}{|\nabla u(x_k)|} - e_N \right) \right| \leq \frac{4\bar{C}^k l^2}{R_0},
$$

provided that $|\mathfrak{x} - x_k| \leq 4l$ (and observe that this condition is fulfilled by any $(\mathfrak{x}, u(\mathfrak{x})) \in \tilde{\Xi}$). By using the latter inequality and the definition of $\tilde{\Xi}$, it also follows that

$$
\begin{aligned}
(\mathfrak{x} - x_k) \cdot e_N &\leq (\mathfrak{x} - x_k) \cdot \frac{\nabla u(x_k)}{|\nabla u(x_k)|} + \left| (\mathfrak{x} - x_k) \cdot \left( \frac{\nabla u(x_k)}{|\nabla u(x_k)|} - e_N \right) \right| \leq \\
\text{(5.38)} \quad &\leq \frac{\hat{C}\bar{C}^k l^2}{4 R_0} + \frac{4\bar{C}^k l^2}{R_0} + H_0(u(\mathfrak{x})) - H_0(u(x_k)),
\end{aligned}
$$

for any $(\mathfrak{x}, u(\mathfrak{x})) \in \tilde{\Xi}$; from (5.38), the definition of $R_0$ and the assumptions of Proposition 5.2, we thus deduce that

$$
\text{(5.39)} \quad \mathfrak{x}_N \leq \frac{\bar{C}^{k+1} \theta}{4} + H_0(u(\mathfrak{x})),
$$

for any $(\mathfrak{x}, u(\mathfrak{x})) \in \tilde{\Xi}$, if $\bar{C}$ is large enough. Therefore, thanks to (5.35) and (5.39), we have that $\tilde{\Xi} \subseteq \mathfrak{D}_{k+1}$. From this and (5.34), we gather (5.33), as desired. This says that the hypotheses of Lemma 4.3 are fulfilled by $D_k$, thus we will freely use such result in what follows.

Let now $E_k$ be as in (4.3). From Lemma 4.1, and taking $\hat{C}$ suitably large, we deduce that, for each $Z \in E_k$ there exists $x = x(Z)$ and $x_k = x_k(Z)$ so that $(x_k, u(x_k)) \in \mathfrak{D}_k$, $|x - x_k| \leq \hat{C}$, $Z = \pi_N(x, u(x))$ and

$$
\text{(5.40)} \quad (x - x_k) \cdot \frac{\nabla u(x_k)}{|\nabla u(x_k)|} \leq H_0(u(x)) - H_0(u(x_k)) + \frac{\text{const}\,\hat{C}}{R_k}.
$$

Thus, from (5.37) and (5.40),

$$
\begin{aligned}
(x - x_k) \cdot e_N &\leq (x - x_k) \cdot \frac{\nabla u(x_k)}{|\nabla u(x_k)|} + \left| (x - x_k) \cdot \left( \frac{\nabla u(x_k)}{|\nabla u(x_k)|} - e_N \right) \right| \leq \\
&\leq H_0(u(x)) - H_0(u(x_k)) + \frac{\text{const}\,\hat{C}}{R_k} + \frac{4\bar{C}^k l^2}{R_0},
\end{aligned}
$$

which implies, thanks to the definition of $R_0$ and the assumptions of Proposition 5.2, that

$$
x_N \leq \bar{C}^{k+1} \theta + H_0(u(x)).
$$

Hence, if $\Xi$ is as defined here above in the statement of Proposition 5.2,

$$
\text{(5.41)} \quad E_k \subseteq \Xi.
$$

Also, by Lemma 4.3,

$$
\text{(5.42)} \quad \mathfrak{L}^N\left( E_k \cap Q_l \right) \geq (1 - (1-c)^k)\, \mathfrak{L}^N(Q_l),
$$

for some $c \in (0,1)$. Thus, the claim in Proposition 5.2 follows from (5.41) and (5.42). □

CHAPTER 6

# Proof of Theorem 1.1

First of all, note that $u$ must attain both positive and negative values thanks to the density estimates in [**28**]. Thus, possibly replacing $l$ by $Cl$, we may assume that $u$ is a local minimizer for $\mathcal{F}$ in $\{|x'| < Cl\} \times \{|x_N| < Cl\}$, that $u(0) = 0$ and that

(6.1) $\qquad u(x) > 0$ if $x_N > \theta > 0$ and $u(x) < 0$ if $x_N < -\theta$.

The strategy for proving Theorem 1.1 consists in assuming, by contradiction, that there exists a point in $\{u = 0\} \cap \{|x'| < l/4\}$ close to $x_N = -\theta$. The contradiction will be, then, that the energy of $u$ is larger than it should.

The first step in proving Theorem 1.1 is thus the following: we assume, by contradicting Theorem 1.1, that

$$\{u = 0\} \cap \{|x'| < \bar{C}^{-k_0} l/4\} \cap \{x_N < (-1 + \bar{C}^{-k_0}/4)\theta\} \neq \emptyset,$$

with $k_0 \in \mathbb{N}$ large and $\frac{\theta}{l}$ small (possibly in dependence of $k_0$). We also set

$$\Xi_0 := \Big\{(x, u(x)) \in \mathbb{R}^N \times \mathbb{R} \text{ s.t.}$$
$$x_N \leq H_0(u(x)) - \theta/2, \ |x' - (x^*)'| \leq l/2, \ |u(x)| \leq 1/2\Big\}.$$

Then, we claim that

(6.2) $\qquad \mathfrak{L}^N\Big(\pi_N(\Xi_0)\Big) \geq (1 - (1 - c_0)^{k_0})\, \mathfrak{L}^N(Q_{l/2})\,,$

for a suitable constant $c_0 > 0$. To prove this, let

$$x^* \in \{u = 0\} \cap \{|x'| < \bar{C}^{-k_0} l/4\} \cap \{x_N < (-1 + \bar{C}^{-k_0}/4)\theta\}$$

Define $\theta^* := \theta/(4\bar{C}^{k_0})$ and $v(x) := u(x + x^*)$. Notice that $v(0) = 0$ and $v(x) < 0$ if $x_N < -\theta^*$. Also $v$ is a local minimizer for $\mathcal{F}$. Then, if we define

$$\Xi^* := \Big\{(z, v(z)) \in \mathbb{R}^N \times \mathbb{R} \text{ s.t.} \quad |z'| \leq l/2$$
$$z_N \leq H_0(v(z)) + \bar{C}^{k_0}\theta^*, \ |v(z)| \leq 1/2\Big\},$$

we deduce by Proposition 5.2 that

$$\mathfrak{L}^N\Big(\pi_N(\Xi^*)\Big) \geq (1 - (1 - c_0)^{k_0})\, \mathfrak{L}^N(Q_{l/2})\,.$$

By elementary computations, one also sees that

$$\Xi^* + (x^*, 0) \subseteq \Xi_0\,,$$

thus proving (6.2).

Let now
$$\Xi_1 := \{(x, u(x)) \in \mathbb{R}^N \times \mathbb{R} \text{ s.t.}$$
$$x_N \geq H_0(u(x)) - \theta/4, \ |x'| \leq l/2, \ |u(x)| \leq 1/2\}.$$

Then, we claim that

(6.3) $$\mathcal{L}^N\left(\pi_N(\Xi_1)\right) \geq c_1 \mathcal{L}^N(Q_{l/2}),$$

for a suitable constant $c_1 > 0$, provided that $\theta/l$ is suitably small. To prove (6.3), let
$$\tilde{u}(x) = \tilde{u}(x', x_N) := -u(x', -x_N),$$
$$\tilde{h}_0(s) : = h_0(-s).$$

Then, $\tilde{h}_0$ satisfies the same assumptions as $h_0$ and $\tilde{u}$ is a local minimizer for the functional
$$\tilde{\mathcal{F}}(v) := \int \frac{|\nabla v|^p}{p} + \tilde{h}_0(v).$$

Hence, we may apply Lemma 5.1 with $h_0$ replaced by $\tilde{h}_0$, and deduce that, if
$$\tilde{\Xi}_1 := \{(x, \tilde{u}(x)) \in \mathbb{R}^N \times \mathbb{R} \text{ s.t.}$$
$$x_N \leq \tilde{H}_0(\tilde{u}(x)) + \theta/4, \ |x'| \leq l/2, \ |u(x)| \leq 1/2\},$$

then
$$\mathcal{L}^N\left(\pi_N(\tilde{\Xi}_1)\right) \geq \text{const } l^{N-1}.$$

From this, (6.3) easily follows.

We now make some remarks on the measure properties of the above sets. First note that, by construction,
$$\pi_N(\Xi_0) \subseteq Q_{\frac{l}{2} + \frac{l}{4\bar{C}k_0}},$$

therefore
$$\mathcal{L}^N\left(\pi_N(\Xi_0) \setminus Q_{l/2}\right) \leq \mathcal{L}^N\left(Q_{\frac{l}{2}+\frac{l}{4\bar{C}k_0}} \setminus Q_{l/2}\right) \leq$$
$$\leq \frac{\text{const } l^{N-1}}{\bar{C}(N-1)k_0} \leq$$
$$\leq \frac{\text{const}}{\bar{C}(N-1)k_0} \mathcal{L}^N(Q_{l/2}).$$

This and (6.2), by assuming $k_0$ large enough, yield that
$$\mathcal{L}^N\left(\pi_N(\Xi_0) \cap Q_{l/2}\right) \geq \left(1 - (1-c_0)^{k_0} - \frac{\text{const}}{\bar{C}(N-1)k_0}\right) \mathcal{L}^N(Q_{l/2}) \geq$$
$$\geq \left(1 - \frac{c_1}{2}\right) \mathcal{L}^N(Q_{l/2}),$$

where $c_1$ is the constant introduced here above. Thus,
$$\mathcal{L}^N\left(Q_{l/2} \setminus \left(\pi_N(\Xi_0) \cap Q_{l/2}\right)\right) \leq \frac{c_1}{2} \mathcal{L}^N(Q_{l/2}).$$

From this, (6.3) and the fact that $\pi_N(\Xi_1) \subseteq Q_{l/2}$, we gather that

$$c_1 \mathcal{L}^N(Q_{l/2}) \leq \mathcal{L}^N\left(\pi_N(\Xi_1)\right) \leq$$
$$\leq \mathcal{L}^N\left(\pi_N(\Xi_1) \cap \left(\pi_N(\Xi_0) \cap Q_{l/2}\right)\right) +$$
$$+ \mathcal{L}^N\left(\pi_N(\Xi_1) \setminus \left(\pi_N(\Xi_0) \cap Q_{l/2}\right)\right) \leq$$
$$\leq \mathcal{L}^N\left(\pi_N(\Xi_0) \cap \pi_N(\Xi_1)\right) +$$
$$+ \mathcal{L}^N\left(Q_{l/2} \setminus \left(\pi_N(\Xi_0) \cap Q_{l/2}\right)\right) \leq$$
$$\leq \mathcal{L}^N\left(\pi_N(\Xi_0) \cap \pi_N(\Xi_1)\right) + \frac{c_1}{2} \mathcal{L}^N(Q_{l/2}),$$

that is

(6.4) $$\mathcal{L}^N\left(\pi_N(\Xi_0) \cap \pi_N(\Xi_1)\right) \geq \frac{c_1}{2} \mathcal{L}^N(Q_{l/2}).$$

On the other hand,

(6.5) $$\Xi_0 \cap \Xi_1 \subseteq \left\{-\frac{\theta}{4} \leq x_N - H_0(u(x)) \leq -\frac{\theta}{2}\right\} = \emptyset.$$

Let now

$$\mathfrak{V} := \left\{Z \in Q_{l/2} \,\Big|\, \exists \tilde{x} \neq \hat{x}, \text{ s.t. } Z = \pi_N(\tilde{x}, u(\tilde{x})) = \pi_N(\hat{x}, u(\hat{x}))\right\}.$$

By (6.5), we have that

$$\mathfrak{V} \supseteq \pi_N(\Xi_0) \cap \pi_N(\Xi_1),$$

thus, due to (6.4),

(6.6) $$\mathcal{L}^N(\mathfrak{V}) \geq \text{const } l^{N-1}.$$

With these inequalities in hand, we now start to estimate the functional, in order to show that the energy of $u$ is too large, and hence obtaining a contradiction.

First of all, for any $x' \in \mathbb{R}^N$ with $|x'| \leq l$, let us define

$$T_{x'}(x_N) := u(x', x_N),$$
$$\mathfrak{C}_{x'} := \{x_N \in \mathbb{R} \mid DT_{x'}(x_N) = 0\}.$$

By standard regularity results (see [15] and [34]), we have that $T_{x'}$ is $C^1$. Hence, by Sard's Lemma,

(6.7) $$\mathcal{L}^N\left(T_{x'}(\mathfrak{C}_{x'})\right) = 0.$$

Thus, using that $T_{x'}$ is locally invertible on the complement of $\mathfrak{C}_{x'}$, we may write the latter set as

$$\bigcup_a J_{a,x'},$$

in such a way $T_{x'}\big|_{J_{a,x'}}$ is a diffeomorphism. Therefore, by Young's inequality (writing $q$ for the dual exponent of $p$) and by changing variable $x_{N+1} := T_{x'}(x_N)$,

$$\int_{J_{a,x'}} \frac{|\nabla u(x', x_N)|^p}{p} + h_0(u(x', x_N))\, dx_N \geq$$
$$\geq \int_{J_{a,x'}} \frac{|\partial_N u(x', x_N)|^p}{p} + h_0(u(x', x_N))\, dx_N =$$
$$= \int_{J_{a,x'}} \frac{|DT_{x'}(x_N)|^p}{p} + h_0(T_{x'}(x_N))\, dx_N \geq$$
$$\geq \int_{J_{a,x'}} \Big(q\, h_0(T_{x'}(x_N))\Big)^{1/q} |DT_{x'}(x_N)|\, dx_N =$$
$$= \int_{T_{x'}(J_{a,x'})} \Big(q\, h_0(x_{N+1})\Big)^{1/q} dx_{N+1},$$

therefore,

$$\sum_a \int_{|x'|\leq l} \int_{T_{x'}(J_{a,x'})} \Big(q\, h_0(x_{N+1})\Big)^{1/q} dx_{N+1}\, dx' \leq$$
$$\leq \sum_a \int_{|x'|\leq l} \int_{J_{a,x'}} \frac{|\nabla u(x', x_N)|^p}{p} + h_0(u(x', x_N))\, dx_N\, dx' =$$
$$= \int_{|x'|\leq l} \int_{[-Cl,Cl]\setminus \mathfrak{C}_{x'}} \frac{|\nabla u(x', x_N)|^p}{p} + h_0(u(x', x_N))\, dx_N\, dx' \leq$$
(6.8)
$$\leq \mathcal{F}_{A_l}(u),$$

where $A_l := \{|x'| < l\} \times \{|x_N| < Cl\}$. Now, we notice that

$$\mathfrak{V} \subseteq \Big\{(x', 0, x_{N+1})\;\Big|\; |x'| \leq l,\; x_{N+1} \in T_{x'}(J_{a,x'}) \cap T_{x'}(J_{\hat{a},x'}) \text{ for some } a \neq \hat{a}\Big\},$$

and that

$$\mathfrak{V} \subseteq Q_l \subseteq \{|x_{N+1}| \leq 1/2\},$$

hence, recalling also (6.6),

$$\int_{|x'|\leq l} \int_{x_{N+1} \in \bigcup_{a\neq \hat{a}} T_{x'}(J_{a,x'}) \cap T_{x'}(J_{\hat{a},x'})} \Big(q\, h_0(x_{N+1})\Big)^{1/q} dx_{N+1}\, dx' \geq$$
$$\geq \int_{\mathfrak{V}} \Big(q\, h_0(x_{N+1})\Big)^{1/q} d(x', x_{N+1}) \geq$$
$$\geq \inf_{[-1/2, 1/2]} (q\, h_0)^{1/q}\, \mathfrak{L}^N(\mathfrak{V}) \geq$$
$$\geq \tilde{c}_1\, l^{N-1} \inf_{[-1/2, 1/2]} (q\, h_0)^{1/q} \geq$$
$$\geq \tilde{c}_1\, l^{N-1} \inf_{[-1/2, 1/2]} (q\, h_0)^{1/q},$$

for a suitably small positive constant $\tilde{c}_1$. Therefore, we gather from the above inequality that

$$\sum_a \int_{|x'|\leq l} \int_{T_{x'}(J_{a,x'})} \left(q\, h_0(x_{N+1})\right)^{1/q} dx_{N+1}\, dx' \geq$$

$$\geq \int_{|x'|\leq l} \int_{\bigcup_a T_{x'}(J_{a,x'})} \left(q\, h_0(x_{N+1})\right)^{1/q} dx_{N+1}\, dx' +$$

$$+ \int_{|x'|\leq l} \int_{x_{N+1}\in \bigcup_{a\neq \hat{a}} T_{x'}(J_{a,x'})\cap T_{x'}(J_{\hat{a},x'})} \left(q\, h_0(x_{N+1})\right)^{1/q} dx_{N+1}\, dx' \geq$$

$$\geq \int_{|x'|\leq l} \int_{x_{N+1}\in u(x',[-Cl,Cl]\setminus \mathfrak{C}_{x'})} \left(q\, h_0(x_{N+1})\right)^{1/q} dx_{N+1}\, dx' +$$

$$+ \tilde{c}_1\, l^{N-1} \inf_{[-1/2,1/2]} (q\, h_0)^{1/q}.$$

Thus, due to (6.7),

$$\sum_a \int_{|x'|\leq l} \int_{T_{x'}(J_{a,x'})} \left(q\, h_0(x_{N+1})\right)^{1/q} dx_{N+1}\, dx' \geq$$

$$\geq \int_{|x'|\leq l} \int_{x_{N+1}\in u(x',[-l,l])} \left(q\, h_0(x_{N+1})\right)^{1/q} dx_{N+1}\, dx' +$$

(6.9) $$+ \tilde{c}_1\, l^{N-1} \inf_{[-1/2,1/2]} (q\, h_0)^{1/q}.$$

On the other hand, from Corollary 2.24, we get that

$$\int_{|x'|\leq l} \int_{x_{N+1}\in u(x',[-l,l])} \left(q\, h_0(x_{N+1})\right)^{1/q} dx_{N+1}\, dx' \geq$$

$$\geq \int_{|x'|\leq l} \int_{-1+s_l}^{1-s_l} \left(q\, h_0(x_{N+1})\right)^{1/q} dx_{N+1}\, dx' =$$

(6.10) $$= \omega_{N-1}\, l^{N-1} \int_{-1+s_l}^{1-s_l} \left(q\, h_0(x_{N+1})\right)^{1/q} dx_{N+1},$$

where $\omega_{N-1}$, as usual, denotes the volume of the $(N-1)$-dimensional unit ball. From (6.9) and (6.10), we thus obtain that

$$\sum_a \int_{|x'|\leq l} \int_{T_{x'}(J_{a,x'})} \left(q\, h_0(x_{N+1})\right)^{1/q} dx_{N+1}\, dx' \geq$$

$$\geq \omega_{N-1}\, l^{N-1} \int_{-1+s_l}^{1-s_l} \left(q\, h_0(x_{N+1})\right)^{1/q} dx_{N+1} +$$

$$+ \tilde{c}_1\, l^{N-1} \inf_{[-1/2,1/2]} (q\, h_0)^{1/q},$$

and, therefore, thanks to (6.8),

$$\mathcal{F}_{A_l}(u) \geq$$

$$\geq \omega_{N-1}\, l^{N-1} \int_{-1+s_l}^{1-s_l} \left(q\, h_0(x_{N+1})\right)^{1/q} dx_{N+1} +$$

(6.11) $$+ \tilde{c}_1\, l^{N-1} \inf_{[-1/2,1/2]} (q\, h_0)^{1/q}.$$

We now notice that, if $c_2$ and $c_3$ are positive constants, suitably small with respect to $\tilde{c}_1$, one has

$$\omega_{N-1} l^{N-1} \int_{[-1,-1+c_2]\cup[1-c_2,1]} \left(q h_0(x_{N+1})\right)^{1/q} dx_{N+1} \leq$$
$$\leq 2c_2 \omega_{N-1} l^{N-1} \sup_{[-1,1]} (q h_0(x_{N+1}))^{1/q} \leq$$
(6.12) $$\leq \frac{\tilde{c}_1}{2} l^{N-1} \inf_{[-1/2,1/2]} (q h_0)^{1/q}$$

and

(6.13) $$c_3 l^{N-1} \leq \frac{\tilde{c}_1}{2} l^{N-1} \inf_{[-1/2,1/2]} (q h_0)^{1/q}.$$

We now assume $l$ big enough so that $s_l < c_2$: then, by means of (6.11), (6.12) and (6.13),

$$\mathcal{F}_{A_l}(u) \geq \omega_{N-1} l^{N-1} \int_{-1}^{1} \left(q h_0(x_{N+1})\right)^{1/q} dx_{N+1} +$$
(6.14) $$+ c_3 l^{N-1}.$$

This estimate will say that the energy of $u$ is too large (thanks to the term "$c_3 l^{N-1}$" here above), and it will provide the desired contradiction. For this, let us define the rescaled functional

$$\mathcal{F}_\Omega^\varepsilon(v) := \int_\Omega \frac{\varepsilon^{p-1}|\nabla v(x)|^p}{p} + \frac{1}{\varepsilon} h_0(v(x)) \, dx.$$

Then, if $\varepsilon := 1/l$ and $u_\varepsilon(x) := u(x/\varepsilon)$, by scaling (6.14), we deduce that

$$\mathcal{F}_{A_1}^\varepsilon(u_\varepsilon) = \varepsilon^{N-1} \mathcal{F}_{A_l}(u) \geq$$
(6.15) $$\geq \omega_{N-1} \int_{-1}^{1} \left(q h_0(x_{N+1})\right)^{1/q} dx_{N+1} + c_3.$$

On the other hand, by §3 of [7], up to subsequences, we have that $u_\varepsilon$ converges almost everywhere and in $L^1_{\text{loc}}$ to the step function $\chi_E - \chi_{\mathbb{R}^N \setminus E}$, for a suitable set $E \subseteq \mathbb{R}^N$, and that

(6.16) $$\lim_{\varepsilon \to 0^+} \mathcal{F}_{A_1}^\varepsilon(u_\varepsilon) = \text{Per}(E, A_1) \int_{-1}^{1} \left(q h_0\right)^{1/q},$$

where, given $A \subseteq B$, we denote the perimeter of $A$ in $B$ as $\text{Per}(A, B)$ (see, for instance, [20] for full details on such definition). As a matter of fact, in our situation, the set $E$ may be better specified, in the following way. From (6.1), there exists $\kappa > 0$ so that $u(x) \geq \kappa$ if $|x'| \leq l$ and $x_N \geq 2\theta$ and $u(x) \leq -\kappa$ if $|x'| \leq l$ and $x_N \leq -2\theta$.

Therefore, $u_\varepsilon(x) \geq \kappa$ if $|x'| \leq 1$ and $x_N \geq 2\varepsilon\theta$ and $u(x) \leq -\kappa$ if $|x'| \leq 1$ and $x_N \leq -2\varepsilon\theta$. In particular, for almost any $x \in A_1$,

$$\lim_{\varepsilon \to 0^+} u_\varepsilon(x) \geq \kappa \text{ if } x_N > 0 \text{ and}$$
$$\lim_{\varepsilon \to 0^+} u_\varepsilon(x) \leq -\kappa \text{ if } x_N > 0.$$

This implies that $E = A_1 \cap \{x_N > 0\}$. And so $\text{Per}(E, A_1) = \omega_{N-1}$. Therefore, from (6.16)
$$\lim_{\varepsilon \to 0^+} \mathcal{F}_{A_1}^\varepsilon(u_\varepsilon) = \omega_{N-1} \int_{-1}^{1} \left(q\, h_0\right)^{1/q}.$$
This contradicts (6.15) and finishes the proof of Theorem 1.1.

CHAPTER 7

# Proof of Theorem 1.2

The proof of Theorem 1.2 will be performed by compactness, by using Theorem 1.1 and a result of [30].

We fix $\theta_0 > 0$ and we assume by contradiction that there exist $u_k$, $\theta_k$, $l_k$ for which

**(C1)** $u_k$ is a local minimizer for $\mathcal{F}$ in $\{|x'| < l_k\} \times \{|x_N| < l_k\}$, with $u_k(0) = 0$.

**(C2)** $\{u_k = 0\} \subseteq \{|x'| < l_k\} \times \{|x_N| < \theta_k\}$, with $\theta_k \geq \theta_0$ and $\frac{\theta_k}{l_k} \longrightarrow 0$ when $k \to \infty$,

but the thesis of Theorem 1.2 does not hold. Let us consider the following rescaling:

$$(7.1) \qquad y' = \frac{x'}{l_k} \quad ; \quad y_N = \frac{x_N}{\theta_k}$$

say $(y', y_N) = T(x', x_N)$. Define

$$A_k := \{(y', y_N) \text{ s.t. } T^{-1}(y', y_N) \in \{u_k = 0\}\} = T\Big(\{u_k = 0\}\Big).$$

**STEP 1:** There exists a Hölder continuous function $w : \mathbb{R}^{N-1} \to \mathbb{R}$ such that: if we define

$$A_\infty := \Big\{(y', w(y')), |y'| \leq \frac{1}{2}\Big\}$$

then, for any $\varepsilon > 0$, $A_k \cap \{|y'| \leq 1/2\}$ lies in a $\varepsilon$-neighborhood of $A_\infty$, for $k$ sufficiently large.

*Proof of step 1.*
Let us suppose that

$$y_0 = (y'_0, y_{0\,N}) \in A_k \qquad \text{with} \qquad |y'_0| \leq 1/2.$$

Then, $u_k(l_k y'_0, \theta_k y_{0\,N}) = 0$, and so, by means of **(C2)**, $|\theta_k y_{0\,N}| < \theta_k$; therefore, using again **(C2)**, we infer that

$$\{u_k = 0\} \subseteq \{|x_N - \theta_k y_{0\,N}| < 2\theta_k\}.$$

Thence, we can exploit Theorem 1.1 in the cylinder

$$(7.2) \qquad \begin{aligned} \{|x' - l_k y'_0| < \frac{l_k}{2}\} \times \{|x_N - \theta_k y_{0\,N}| < 2\theta_k\} &\subseteq \\ \subseteq \{|x'| < l_k\} \times \{|x_N| < l_k\}, \end{aligned}$$

and get that there exists a universal constant $\eta_0 > 0$ such that

$$\{u_k = 0\} \cap \{|x' - l_k y'_0| < \eta_0 \frac{l_k}{2}\} \subseteq \{|x_N - \theta_k y_{0\,N}| < 2(1 - \eta_0)\theta_k\},$$

provided
$$\frac{4\theta_k}{l_k} \leq \varepsilon_0(2\theta_0),$$
where $\varepsilon_0(\cdot)$ is the one given by Theorem 1.1. Rescaling back, we get
$$A_k \cap \{|y' - y_0'| < \frac{\eta_0}{2}\} \subseteq \{|y_N - y_{0\,N}| < 2(1 - \eta_0)\}.$$
By iterating, we get
(7.3) $$A_k \cap \{|y' - y_0'| < \frac{\eta_0^m}{2}\} \subseteq \{|y_N - y_{0\,N}| < 2(1 - \eta_0)^m\},$$
provided
(7.4) $$\frac{4\theta_k}{l_k} \leq \eta_0^{m-1}\varepsilon_0\left(2(1 - \eta_0)^{m-1}\theta_0\right).$$

We now fix $m_0 \in \mathbb{N}$ and consider $m \leq m_0$ (later on, during a limiting procedure performed on page 71, we let $m_0 \longrightarrow +\infty$). Note that, in this setting, (7.4) (and therefore (7.3)) is fulfilled for $k$ suitably large, say $k \geq k^\star(m_0)$. We claim that $A_k \cap \{|y'| \leq 1/2\}$ is above the graph of
(7.5) $$\Psi_{y_0,k}(y') = y_{0\,N} - 2(1 - \eta_0)^{m_0} - \alpha|y' - y_0'|^\beta$$
where $\alpha$ and $\beta > 0$ depend only on $\eta_0$.

To prove this, let $(y', y_N) \in A_k \cap \{|y'| \leq 1/2\}$. Since $|y_0'| \leq \frac{1}{2}$ we have that $|y' - y_0'| \leq 1$. Now, we consider three different cases: the case $|y' - y_0'| \leq \frac{\eta_0^{m_0}}{2}$, the case $\frac{\eta_0^{m_0}}{2} \leq |y' - y_0'| \leq \frac{1}{2}$, and the case $\frac{1}{2} \leq |y' - y_0'| \leq 1$.

In case $|y' - y_0'| \leq \frac{\eta_0^{m_0}}{2}$, (7.5) follows immediately from (7.3), with $m = m_0$. If, on the other hand, $\frac{\eta_0^{m_0}}{2} \leq |y' - y_0'| \leq \frac{1}{2}$, then we argue as follows. We first note that, in this case, there exists $m$ with $0 \leq m \leq m_0$, such that
(7.6) $$\frac{\eta_0^{m+1}}{2} \leq |y' - y_0'| \leq \frac{\eta_0^m}{2}.$$

Consequently, from (7.3), we have that
(7.7) $$2(1 - \eta_0)^m \geq |y_N - y_{0\,N}|$$

By (7.6) and the fact that $0 < \eta_0 < 1$, we also get
$$m \leq \frac{-\ln(2|y' - y_0'|)}{\ln(\frac{1}{\eta_0})} \leq m + 1.$$

In particular, it follows that
$$(1 - \eta_0)^m \leq (1 - \eta_0)^{\left(\frac{-\ln(2|y' - y_0'|)}{\ln(\frac{1}{\eta_0})} - 1\right)} =$$
$$= \frac{1}{(1 - \eta_0)} e^{\beta \ln(2|y' - y_0'|)} = \frac{(2|y' - y_0'|)^\beta}{(1 - \eta_0)},$$
where $\beta := \frac{-\ln(1 - \eta_0)}{\ln(\frac{1}{\eta_0})}$.

Therefore, recalling (7.7), it follows
$$|y_N - y_{0\,N}| \leq \frac{2^{\beta+1}}{(1 - \eta_0)}|y' - y_0'|^\beta$$

which is the desired result, with $\alpha := 2^{\beta+1}/(1-\eta_0)$.

Finally, eventually adding[1] a constant to $\alpha$, the result also follows for the case $|y' - y'_0| \in [1/2, 1]$. This ends the proof of (7.5).

Note now that, as $y_0$ varies, $\Psi_{y_0,k}$ are Hölder continuous functions with Hölder modulus of continuity bounded via the function $\alpha t^\beta$ (recall that $m_0$ is fixed for the moment, and that $\alpha$ and $\beta$ depend only on $\eta_0$). Therefore, if we set

$$\psi_k(y') := \sup_{\substack{|y'_0| \leq \frac{1}{2} \\ y_0 \in A_k}} \Psi_{y_0,k}(y')$$

then, $\psi_k$ is a Hölder continuous function (with Hölder modulus of continuity bounded via the function $\alpha t^\beta$), and $A_k \cap \{|y'| \leq 1/2\}$ is above the graph of $\psi_k$.

Arguing in the same way, possibly taking $\alpha$ and $\beta$ larger (depending only on $\eta_0$), we also get that, if we define

$$\Phi_{y_0,k}(y') := y_{0\,N} + 2(1-\eta_0)^{m_0} + \alpha|y' - y'_0|^\beta,$$

then $A_k \cap \{|y'| \leq 1/2\}$ is below the graph of $\Phi_{y_0,k}$. Arguing as above, we define

$$\phi_k(y') := \inf_{\substack{|y'_0| \leq \frac{1}{2} \\ y_0 \in A_k}} \Phi_{y_0,k}(y'),$$

so that $\phi_k$ is a Hölder continuous function (with Hölder modulus of continuity bounded via the function $\alpha t^\beta$), and $A_k \cap \{|y'| \leq 1/2\}$ is below the graph of $\phi_k$.

In particular, $A_k \cap \{|y'| \leq 1/2\}$ lies between the graphs of $\psi_k(y')$ and $\phi_k(y')$ for any $k \geq k^\star(m_0)$ and, by construction,

(7.8) $$0 \leq \phi_k(y') - \psi_k(y') \leq 4(1-\eta_0)^{m_0}.$$

Also, for $m_0$ fixed, by Ascoli-Arzelà Theorem, letting $k \to \infty$, it follows that $\psi_k(y')$ uniformly converges to a Hölder continuous function which depends on $m_0$, say

$$\lim_{k \to +\infty} \psi_k(y') \to w^-_{m_0}(y').$$

Analogously, we find a Hölder continuous function $w^+_{m_0}$, such that

$$\lim_{k \to +\infty} \phi_k(y') \to w^+_{m_0}(y')$$

uniformly. Also, by construction, we have that $w^-_{m_0} \leq w^+_{m_0}$ and that

(7.9) $$A_k \cap \{|y'| \leq 1/2\} \text{ lies between}$$
$$\text{the graphs of } w^-_{m_0} - \varepsilon/2 \text{ and } w^+_{m_0} + \varepsilon/2,$$

for $k$ large.

Let now $m_0 \to \infty$. In this case, by Ascoli-Arzelà Theorem[2], we get that there exists a Hölder continuous function $w$ such that $w^-_{m_0}$ uniformly converges to $w$. By (7.8), also $w^+_{m_0}$ uniformly converges to $w$. The claim thus follows from (7.9).

**STEP 2:** The function $w$ constructed in the first step is harmonic.

---

[1] Notice indeed that, by **(C2)**, we have that
$$|y_N - y_{0\,N}| \leq |y_N| + |y_{0\,N}| \leq 2.$$

[2] We remark that, by the construction of $\alpha$ and $\beta$ above, the Hölder constants of $w^\pm_{m_0}$ depend on $\eta_0$, but are independent of $m_0$.

*Proof of step 2.*
We prove that $w$ is harmonic in the viscosity sense. Then it follows that it is harmonic in the classic sense (see, e.g., Theorem 6.6 in [8]).
For this, let $P$ be the quadratic polynomial

$$P(y') := \frac{1}{2} y'^T M y' + \xi \cdot y'.$$

Assume, by contradiction, that $\Delta P > 0$, that $P$ touches the graph of $w$, say at $0$ for simplicity and that $P$ stays below it in $|y'| < 2r$, for some $r \in (0,1)$. Let now $\delta_0 > 0$ be the universal constant of Lemma 9.3 in [30] and let us define

$$\delta := \min\left\{ \left(\frac{\Delta P}{2\theta_0}\right)^{\frac{1}{2}}, \frac{1}{2\theta_0 \|M\|}, \frac{1}{2\theta_0 |\xi|}, \left(\frac{\delta_0}{2\theta_0}\right)^{\frac{1}{2}}, r \right\}.$$

Thus, $\delta$ is such that

$$\Delta P > 2\delta^2 \theta_0, \quad \|M\| \leq \frac{1}{2\delta\theta_0}, \quad |\xi| \leq \frac{1}{2\delta\theta_0},$$

(7.10)
$$\delta^2 \theta_0 \leq \frac{\delta_0}{2}.$$

Note that, eventually replacing $\delta$ with $2\delta$ and $P(y')$ with $P(y') - \delta|y'|^2$, we may assume, with no lose of generality, that $P$ touches the graph of $w$ at $0$ and stays strictly below it in $|y'| < 2\delta < 2$. therefore, since $A_k \cap \{|y'| \leq 1/2\}$ uniformly converges to the graph of $w$, it follows that, for $k$ large, we find points $y_k = (y'_k, y_{k\,N})$ close to $0$, such that $P(y') - K_k$ touches $A_k$ at $(y'_k, y_{k\,N})$ and stays below it in $|y' - y'_k| \leq \delta$, for an appropriate $K_k \in \mathbb{R}$. In particular, we have

(7.11)
$$y_{k\,N} + K_k = \frac{1}{2} y'^T_k M y'_k + \xi \cdot y'_k.$$

Let us now consider the following translation

$$z' = y' - y'_k \qquad z_N = y_N - (y_{k\,N} + K_k)$$

Exploiting (7.11) we find a surface

$$\left\{ z_N = \frac{1}{2} z'^T M z' + \xi_k \cdot z' \right\},$$

with

$$\xi_k := M y'_k + \xi$$

that touches $A_k$ by below at the origin and stays below it in $|z'| < \delta$. Notice also that, by construction,

(7.12)
$$|\xi_k| \leq \frac{1}{\delta\theta_0}.$$

Rescaling back, we get that the surface

$$\left\{ x_N = \frac{\theta_k}{l_k^2} \frac{1}{2} x'^T M x' + \frac{\theta_k}{l_k} \xi_k \cdot x' \right\}$$

touches $\{u_k = 0\}$ at the origin and stays below it, if $|x'| < \delta l_k$.
We write now the above surface in the form

$$\left\{ x_N = \frac{\delta^2 \theta_k}{(\delta l_k)^2} \frac{1}{2} x'^T M x' + \frac{\delta^2 \theta_k}{\delta l_k} \frac{1}{\delta} \xi_k \cdot x' \right\}$$

# 7. PROOF OF THEOREM 1.2

and we exploit[3] Lemma 9.3 in [**30**], thus gathering that
$$\Delta P \leq \delta^2 \theta_0,$$
against the assumption. This contradiction shows that $\Delta P \leq 0$. By arguing in the same way, one may prove that $\Delta P \geq 0$ if $P$ touches $w$ by above, so that the claim of Step 2 on page 71 is proved.

**CONCLUSION:** Since $w$ is harmonic, by standard elliptic estimates (see, e.g., Theorem 2.10 in [**19**]), we find a positive universal constant $C$, such that
$$\|D^2 w\| \leq C$$
Therefore, since by construction $w(0) = 0$, by Taylor's formula, it follows that
$$|w(y') - \nabla w(0) \cdot y'| < C' \eta_2^2 \quad \text{for} \quad |y'| < 2\eta_2.$$
In particular, for $\eta_2$ sufficiently small, setting
$$\xi' := \nabla w(0),$$
we get that there exist positive constants $0 < \eta_1 < \eta_2 < 1$, for which
$$(7.13) \qquad |w(y') - \xi' \cdot y'| < \frac{\eta_1}{2} \quad \text{for} \quad |y'| < 2\eta_2.$$
Now, let us consider
$$(7.14) \qquad \xi_k := \frac{\left(\frac{\theta_k}{l_k}\xi', -1\right)}{\sqrt{\frac{\theta_k^2}{l_k^2}|\xi'|^2 + 1}}.$$
Considering the rescaling given by (7.1), elementary geometric considerations show that
$$(7.15) \qquad \{|\pi_{\xi_k} x| < \eta_2 l_k\} \times \{|x \cdot \xi_k| < \eta_2 l_k\} \subset \{|x'| < 2 l_k \eta_2\} \subset \{|x'| < l_k/2\}.$$
Since $A_k \cap \{|y'| \leq 1/2\}$ uniformly converges to the graph of $w$, for $k$ sufficiently large (thanks to Step 1 on page 69), we may suppose that $A_k \cap \{|y'| \leq 1/2\}$ is in a $\frac{\eta_1}{4}$-neighborhood of the graph of $w$. Consequently, by (7.13), taking into account the rescaling, it follows that
$$\{u_k = 0\} \cap \{|x'| \leq l_k/2\} \subset \left\{\left|x_N - \frac{\theta_k}{l_k}\xi' \cdot x'\right| < \frac{3}{4}\theta_k \eta_1\right\}.$$
From (7.14), we thence get that
$$\{u_k = 0\} \cap \{|x'| \leq l_k/2\} \subset \left\{|x \cdot \xi_k| < \frac{3}{4}\theta_k \eta_1\right\},$$
which, together with (7.15), is a contradiction with the fact that $u_k$ does not satisfies the statement of Theorem 1.2. This ends the proof of Theorem 1.2.

---

[3]Notation remark: Lemma 9.3 in [**30**] is used here with $M_1 := M$, $\delta := \delta^2 \theta_0$, $\theta := \delta^2 \theta_k$ (note that $\delta \leq \theta$ since $\theta_0 \leq \theta_k$), $l := \delta l_k$, and $\xi := \frac{1}{\delta} \xi_k$ (therefore $|\xi| \leq \frac{1}{\delta^2 \theta_0}$, thanks to (7.12)). In particular, with this setting, since $\frac{\theta_k}{l_k} \to 0$, then $\frac{\theta}{l} \to 0$. Also, since $\sigma(\delta) > 0$ (where $\sigma(\delta)$ is as in Lemma 9.3 in [**30**]), then the condition $\frac{\theta}{l} < \sigma(\delta)$ (i.e. $\frac{\delta \theta_k}{l_k} < \sigma(\delta)$) is fulfilled for $k$ sufficiently large.

CHAPTER 8

# Proof of Theorem 1.3

The following Lemma 8.1 is an intermediate step towards the proof of Theorem 1.3 and it is also useful for the proof of Theorem 1.4. The proof of Lemma 8.1 is based on an iteration of Theorem 1.2.

LEMMA 8.1. *Let $u$ be a Class A minimizer for $\mathcal{F}$ in $\mathbb{R}^N$ with $u(0) = 0$. Suppose that there exist sequences of positive numbers $\theta_k$, $l_k$ and unit vectors $\xi_k$, with*

(8.1) $$l_k \to \infty \quad \text{and} \quad \frac{\theta_k}{l_k} \to 0,$$

*such that*

(8.2) $$\{u = 0\} \cap \left( \{|\pi_{\xi_k} x| < l_k\} \times \{|x \cdot \xi_k| < l_k\} \right) \subset \{|x \cdot \xi_k| \leq \theta_k\}.$$

*Then, the 0 level set $\{u = 0\}$ is a hyperplane in $\mathbb{R}^N$.*

PROOF. Let fix $\theta_0 > 0$ and $\varepsilon \leq \varepsilon_1(\theta_0)$, with $\varepsilon_1(\theta_0)$ given by Theorem 1.2. We consider $k$ so large in that

(8.3) $$\frac{\theta_k}{l_k} \leq \varepsilon \leq \varepsilon_1(\theta_0).$$

Two cases are now possible: either, for infinitely many $k$'s $\theta_k \leq \theta_0$, or for infinitely many $k$'s $\theta_k > \theta_0$.

In the first case, we take the subsequence of $k$'s for which $\theta_k \leq \theta_0$ and we assume, by possibly extracting a further subsequence, that $\xi_k$ converges to a suitable unit vector $\xi$. We consider a $y$-frame of coordinates in which $y_N$ is parallel to $\xi$. Consequently, by (8.1) and (8.2), we deduce that, in this system of coordinates,

$$\{u = 0\} \subseteq \{|y_N| \leq \theta_0\}.$$

Thence, since $\theta_0$ is arbitrary,

$$\{u = 0\} \subseteq \{y_N = 0\},$$

which proves the desired result.

If, on the other hand, $\theta_k > \theta_0$ for infinitely many $k$'s, then we fix $k$ large enough to fulfill (8.3) and we apply Theorem 1.2 repeatedly as much as we can. More precisely, for $h \geq 0$, let $l_k^{(h)} := \eta_2^h l_k$ and $\theta_k^{(h)} := \eta_1^h \theta_k$. Then, if $\theta_k^{(h)} > \theta_0$, we can keep applying Theorem 1.2; we stop this procedure when $h$ is so large that

$$\theta_k^{(h)} \leq \theta_0.$$

More precisely, we stop the iterative application of Theorem 1.2 when $h \geq 1$ is so that

$$\theta_k^{(h-1)} > \theta_0 \geq \theta_k^{(h)}.$$

For such $h$, we get, by construction, that
$$\theta_0 \geq \theta_k^{(h)} \geq \eta_1 \theta_0.$$
Also, by construction,
$$\frac{\theta_k^{(h)}}{l_k^{(h)}} = \left(\frac{\eta_1}{\eta_2}\right)^h \frac{\theta_k}{l_k} \leq \varepsilon.$$
In particular,
(8.4) $$l_k^{(h)} \geq \frac{\eta_1 \theta_0}{\varepsilon}.$$
What is more, the repeated use of Theorem 1.2, has driven us to proving that, in some system of coordinates,
$$\{u = 0\} \cap \left(\{|y'| < l_k^{(h)}\} \times \{|y_N| < l_k^{(h)}\}\right) \subseteq \{|y_N| \leq \theta_k^{(h)}\},$$
that is,
$$\{u = 0\} \cap \left(\left\{|y'| < \frac{\eta_1 \theta_0}{\varepsilon}\right\} \times \left\{|y_N| < \frac{\eta_1 \theta_0}{\varepsilon}\right\}\right) \subseteq \{|y_N| \leq \theta_0\},$$
thanks to to (8.4). Therefore, letting $\varepsilon \longrightarrow 0$, it follows that
$$\{u = 0\} \subset \{|y_N| \leq \theta_0\}.$$
Since $\theta_0$ was arbitrary, the lemma is proved. $\square$

By means of Lemma 8.1, we are now in the position of completing the proof of Theorem 1.3, by arguing as follows.

Let us consider the rescaled functional
$$\mathcal{F}_\Omega^\varepsilon(v) := \int_\Omega \frac{\varepsilon^{p-1}|\nabla v(x)|^p}{p} + \frac{1}{\varepsilon} h_0(v(x)) \, dx.$$
Then, for any $\Omega \subset \mathbb{R}^N$, $u_\varepsilon(x) := u(x/\varepsilon)$ is a local minimizer for $\mathcal{F}_\Omega^\varepsilon$.

Therefore, by §3 of [7], up to subsequences, we have that $u_\varepsilon$ converges almost everywhere and in $L^1_{\text{loc}}$ to the step function $\chi_E - \chi_{\mathbb{R}^N \setminus E}$, for a suitable set $E \subseteq \mathbb{R}^N$ with minimal perimeter.

We claim now that
$$\{u_{\varepsilon_k} = 0\} \quad \text{uniformly converges to} \quad \partial E \quad \text{on compact sets}.$$
Assume that this is not true and note that in this case there exist $\delta > 0$, and a point $z_0 \in \mathbb{R}^N$ and points $x_k$, such that
$$x_k \in \{u_{\varepsilon_k} = 0\} \cap B(z_0, \delta) \quad \text{with} \quad B(z_0, 2\delta) \cap \partial E = \emptyset$$
Assume e.g. $B(z_0, 2\delta) \subset E$ and note that in this case, exploiting the density estimate in [28], we get a contradiction with the fact that $u_\varepsilon$ converges almost everywhere and in $L^1_{\text{loc}}$ to the step function $\chi_E - \chi_{\mathbb{R}^N \setminus E}$ (in the same way we get a contradiction if $B(z_0, 2\delta) \subset \mathbb{R}^N \setminus E$).

Since $\partial E$ is a minimal surface in $\mathbb{R}^N$, and we assumed that $N \leq 7$, then $\partial E$ is a hyperplane (see, for instance, Theorem 17.3 in [20]). Also, since $u_{\varepsilon_k}(0) = 0$ and $\{u_{\varepsilon_k} = 0\}$ uniformly converges to $\partial E$, it follows that $0 \in \partial E$.
This implies that, in some system of coordinates
(8.5) $$\{u_{\varepsilon_k} = 0\} \cap B_1 \subset \{|x_N| \leq \delta_k\}$$

with $\delta_k \to 0$. Rescaling back we get that

$$\{u=0\} \cap B_{\frac{1}{\varepsilon_k}} \subset \{|x_N| \leq \frac{\delta_k}{\varepsilon_k}\} \tag{8.6}$$

Two cases are now possible: either $\delta_k/\varepsilon_k$ is bounded away from zero, or, up to subsequences, $\delta_k/\varepsilon_k \longrightarrow 0$. In the latter case, we pass to the limit (8.5) by sending $k \longrightarrow +\infty$, getting that $\{u=0\}$ is a hyperplane. If, on the other hand, $\delta_k/\varepsilon_k \geq \theta_0$, for some $\theta_0 > 0$, we define

$$l_k := \frac{1}{2\varepsilon_k}, \qquad \theta_k := \frac{\delta_k}{\varepsilon_k}$$

and we observe that

$$\frac{\theta_k}{l_k} = \frac{\delta_k}{2} \longrightarrow 0;$$

then, it follows that the assumptions of Lemma 8.1 are fulfilled. Thus, the application of Lemma 8.1 proves that $\{u=0\}$ is a hyperplane, which is the desired result.

CHAPTER 9

# Proof of Theorem 1.4

First, we prove the minimality of $u$:

LEMMA 9.1. *Let $h_0$ satisfy (1.1), (1.2), (1.3) and (1.4). Let $u$ be a weak Sobolev solution of (1.5) in the whole $\mathbb{R}^N$, satisfying $|u| \leq 1$, $\partial_N u > 0$ and $\lim_{x_N \to +\infty} u = \pm 1$. Then, $u$ is a class A minimizer.*

PROOF. Since $u$ is strictly increasing, $|u| < 1$. Let $B \subset \mathbb{R}^N$ be a closed ball and let $v$ be a minimizer for $\mathcal{F}_B$ with $v = u$ on $\partial B$. Our aim is to show that $u = v$ in $B$. Let us argue by contradiction and assume, say, that

$$(9.1) \qquad v(x^\star) > u(x^\star),$$

for some $x^\star \in B$. Possibly cutting $v$ on the $\pm 1$-levels (which decreases $\mathcal{F}_B$), we may and do assume that $|v| \leq 1$. More precisely, as mentioned in the footnote on page 2, by (1.4) it follows that $|v| < 1$.
Then, since $|\nabla u| \leq$ const thanks to [**15**] or [**34**], and $\lim_{x_N \to +\infty} u = \pm 1$, we deduce that

$$(9.2) \qquad u(x + t e_N) \geq v(x)$$

for any $x \in B$, provided that $t$ is large enough. Indeed, to prove (9.2), let us argue by contradiction and assume that $u(x_t + t e_N) < v(x_t)$ for some $x_t \in B$ and a diverging sequence of $t$; let also $\alpha > 0$ so that $v \leq 1 - \alpha$. Then, up to subsequence, we may assume that $x_t$ converges to $x_\infty \in B$; but then

$$\begin{aligned}
1 &= \lim_{t \to +\infty} u(x_\infty + t e_N) \leq \\
&\leq \lim_{t \to +\infty} u(x_t + t e_N) + \text{const} \, |x_t - x_\infty| = \\
&= \lim_{t \to +\infty} u(x_t + t e_N) \leq \\
&\leq \lim_{t \to +\infty} v(x_t) \leq \\
&\leq 1 - \alpha.
\end{aligned}$$

This contradiction proves (9.2).

Thanks to (9.2), we thus slide $u(\cdot + t e_N)$ towards the $e_N$-direction until we touch $v$ from above. Say this happen at $\bar{x} \in B$ for $t = \bar{t}$. In the light of (9.1), we have that

$$\begin{aligned}
u(x^\star + \bar{t} e_N) &\geq v(x^\star) > \\
&> u(x^\star),
\end{aligned}$$

thence, since $u$ is strictly increasing in the $e_N$-direction,

$$\bar{t} > 0.$$

Since now $\partial_N u > 0$ we have that $\nabla u(\cdot + \bar{t}e_N) \neq 0$. Therefore, it follows that the assumptions of the Strong Comparison Principle for $p$-Laplace equations in [9] (see Corollary B.5 here) applies to $u(\cdot + \bar{t}e_N)$ and $v$ and so this touching point must occur on $\partial B$, that is $\bar{x} \in \partial B$. Since $u = v$ on $\partial B$, it follows that $v(\bar{x}) = u(\bar{x})$. Consequently, since $u$ is strictly increasing in the $e_N$-direction,

$$u(\bar{x}) = v(\bar{x}) = u(\bar{x} + \bar{t}e_N) > u(\bar{x}).$$

This contradiction shows that (9.1) cannot hold, hence $v \leq u$. Analogously, one sees that $v \geq u$, thence $v = u$. $\square$

By means of Lemma 9.1, we can complete the proof of Theorem 1.4, by arguing as follows. With no loss of generality, we assume that $u(0) = 0$. Then, for $\varepsilon > 0$, setting $u_\varepsilon(x) := u(x/\varepsilon)$, we know from Lemma 9.1 and the results of [7] that $u_\varepsilon$ $L^1_{\text{loc}}$-converges (and thus a.e.-converges), up to subsequence, to $\chi_E - \chi_{\mathbb{R}^N \setminus E}$, for a suitable $E$ with minimal perimeter.

Since $\partial_N u > 0$ and $\lim_{x_N \to +\infty} u = \pm 1$, we have that the zero level set of $u$ is a graph in the $e_N$-direction; more precisely, there exists $\gamma : \mathbb{R}^{N-1} \longrightarrow \mathbb{R}$ so that

$$\{u < 0\} = \{x_N < \gamma(x')\}.$$

By scaling, we thus deduce that

(9.3) $$\{u_\varepsilon < 0\} = \{x_N < \gamma_\varepsilon(x')\},$$

with

$$\gamma_\varepsilon(x') := \varepsilon \gamma(x'/\varepsilon).$$

We now claim that

(9.4) $$\chi_{\{x_N < \gamma_\varepsilon(x')\}} \text{ converges in } L^1_{\text{loc}} \text{ to } \chi_{\mathbb{R}^N \setminus E}.$$

Indeed: we know that $u_\varepsilon$ converges to $\chi_E - \chi_{\mathbb{R}^N \setminus E}$ in $\mathbb{R}^N \setminus Z$, for a suitable set $Z$ with $\mathcal{L}^N(Z) = 0$; thus, if $x \in E \setminus Z$,

$$\lim_{\varepsilon \to 0^+} u_\varepsilon(x) = \chi_E(x) - \chi_{\mathbb{R}^N \setminus E}(x) = 1,$$

therefore

$$\lim_{\varepsilon \to 0^+} \chi_{\{u_\varepsilon < 0\}}(x) = 0 = \chi_{\mathbb{R}^N \setminus E}(x),$$

while if $x \in (\mathbb{R}^N \setminus E) \setminus Z$ then

$$\lim_{\varepsilon \to 0^+} u_\varepsilon(x) = \chi_E(x) - \chi_{\mathbb{R}^N \setminus E}(x) = -1,$$

therefore

$$\lim_{\varepsilon \to 0^+} \chi_{\{u_\varepsilon < 0\}}(x) = 1 = \chi_{\mathbb{R}^N \setminus E}(x).$$

This shows that $\chi_{\{u_\varepsilon < 0\}}$ converges almost everywhere to $\chi_{\mathbb{R}^N \setminus E}$. Thus, (9.4) follows from the Dominated Convergence Theorem and (9.3).

In the light of (9.4) and Lemma 16.3 of [20], we have that $\mathbb{R}^N \setminus E$ is a subgraph of a measurable function which is the a.e.-limit of $\gamma_\varepsilon$ up to subsequences, and which may attain the values $\pm \infty$; that is, there exists $\gamma_\star : \mathbb{R}^{N-1} \longrightarrow [-\infty, +\infty]$ in such a way that

(9.5) $$\gamma_\star(x) = \lim_{\varepsilon \to 0^+} \gamma_\varepsilon(x)$$

## 9. PROOF OF THEOREM 1.4

for almost any $x$, up to subsequence, and
$$\mathbb{R}^N \setminus E = \{x_N < \gamma_\star(x')\}.$$
Since $\partial E$ is a minimal perimeter, we have that $\gamma_\star$ is a quasi-solution of the minimal surface equation, according to Definition 16.1 of [**20**].

We now prove that $\partial E$ is a hyperplane. We distinguish two cases, according to our hypotheses. If $N \le 8$, we have that $\gamma_\star$ is an entire quasi-solution of the minimal surface equation in a space with dimension less or equal than 7; therefore, by Theorem 17.8 and Remark 17.9 of [**20**], we have that $\partial E = \{x_N = \gamma_\star(x')\}$ is a hyperplane. If, on the other hand, $\{u = 0\}$ has at most linear growth, then
$$\{u = 0\} \subseteq \{|x_N| \le K(|x'| + 1)\},$$
for a suitable $K > 0$. This says that
$$\begin{aligned}\{x_N = \gamma_\varepsilon(x')\} &= \{u_\varepsilon = 0\} \subseteq \\ &\subseteq \{|x_N| \le K(|x'| + \varepsilon)\} \subseteq \\ &\subseteq \{|x_N| \le K(|x'| + 1)\},\end{aligned}$$
i.e., $|\gamma_\varepsilon(x')| \le K(|x'| + 1)$. Thus, by means of (9.5), we gather that
$$(9.6) \qquad |\gamma_\star(x')| \le K(|x'| + 1),$$
thus $\gamma_\star$ is locally bounded. Hence, $\gamma_\star$ is a solution of the minimal surface equation (see [**20**], page 183). Therefore, $\partial E$ is a hyperplane thanks to (9.6) and Theorem 17.6 of [**20**].

In any case, we have proved that $\partial E$ is a hyperplane, thence
$$\partial E = \{\xi \cdot x = 0\},$$
for a suitable $\xi \in \mathbb{R}^N$ with $|\xi| = 1$. Thanks to [**28**], we know that $\{u_\varepsilon = 0\}$ $L^\infty_{\text{loc}}$-converges $\partial E$, hence, for any $k \in \mathbb{N}$, there exists $\varepsilon_k > 0$ as small as we wish, so that
$$B_2 \cap \{u_{\varepsilon_k} = 0\} \subseteq \{|\xi \cdot x| \le 1/k\}.$$
By scaling back the variables, we thence obtain that
$$\{u = 0\} \cap \left(\{|x \cdot \xi| \le 1/\varepsilon_k\} \times \{|\pi_\xi x| \le 1/\varepsilon_k\}\right) \subseteq \{|\xi \cdot x| \le 1/(k\varepsilon_k)\}.$$
We now invoke Lemma 8.1, used here with $\xi_k := \xi$, $l_k := 1/\varepsilon_k$, $\theta_k := 1/(k\varepsilon_k)$, and we infer that $\{u = 0\}$ is a hyperplane. This completes the proof of Theorem 1.4.

# APPENDIX A

# Proof of the measure theoretic results

### A.1. Proof of Lemma 4.1

By the hypotheses of the lemma and (2.41), $\widetilde{\mathbb{S}}(Y,R)$ touches the graph of $u$ by above at $X_0$. Notice also that, in the notation of Lemma 3.3, since $Y = F(X_0, \nu^{\widetilde{\mathbb{S}}(Y,R)}(X_0))$,

$$y = x_0 + \frac{\left(\nu_1^{\widetilde{\mathbb{S}}(Y,R)}(X_0), \ldots, \nu_N^{\widetilde{\mathbb{S}}(Y,R)}(X_0)\right)}{\left|\left(\nu_1^{\widetilde{\mathbb{S}}(Y,R)}(X_0), \ldots, \nu_N^{\widetilde{\mathbb{S}}(Y,R)}(X_0)\right)\right|} \sigma(X_0, \nu^{\widetilde{\mathbb{S}}(Y,R)}(X_0)),$$

and thus

$$\text{(A.1)} \quad |(x_0 - y) \cdot e_N| =$$
$$= \frac{\left|\nu_N^{\widetilde{\mathbb{S}}(Y,R)}(X_0)\right|}{\left|\left(\nu_1^{\widetilde{\mathbb{S}}(Y,R)}(X_0), \ldots, \nu_N^{\widetilde{\mathbb{S}}(Y,R)}(X_0)\right)\right|} \left|\sigma(X_0, \nu^{\widetilde{\mathbb{S}}(Y,R)}(X_0))\right| =$$
$$= \frac{|\partial_N u(x_0)|}{|\nabla u(x_0)|} \bigg[R +$$
$$+ H_0(x_0 \cdot e_{N+1}) - H_0\bigg(x_0 \cdot e_{N+1} +$$
$$+ \omega(X_0, \nu^{\widetilde{\mathbb{S}}(Y,R)}(X_0))\bigg) - \frac{\overline{C}_0}{2R}\omega^2(X_0, \nu^{\widetilde{\mathbb{S}}(Y,R)}(X_0))\bigg].$$

Also, from Lemma 3.3,

$$\left[-\frac{1}{4}, \frac{1}{4}\right] \ni y_{N+1} = x_{N+1} + \omega\big(X_0, \nu^{\widetilde{\mathbb{S}}(Y,R)}(X_0)\big)$$

and so

$$\text{(A.2)} \quad \left|\omega\big(X_0, \nu^{\widetilde{\mathbb{S}}(Y,R)}(X_0)\big)\right| \leq \frac{1}{4} + \frac{1}{2} < 1.$$

Therefore, from (A.1), (A.2) and (4.2),

$$|(x_0 - y) \cdot e_N| \geq \operatorname{const} R,$$

provided that $\overline{C}_0$ is large enough; more precisely, since, from (4.2),

$$\frac{x_0 - y}{|x_0 - y|} \cdot e_N = \frac{\nabla u(x_0)}{|\nabla u(x_0)|} \cdot e_N > 0,$$

we have that

$$(x_0 - y) \cdot e_N \geq \operatorname{const} R.$$

Therefore,

$$\text{(A.3)} \quad x_N - y_N \geq \operatorname{const} R,$$

for any $X \in B_{3a}(X_0)$, if $R$ is large enough.

We now point out that, if $X \in B_{3a}(X_0) \cap \widetilde{\mathbb{S}}(Y,R)$,

(A.4) $$\partial_N g_{\widetilde{\mathbb{S}}(Y,R)}(x) \geq \text{const} > 0.$$

In order to prove the above inequality, first notice that, by a direct computation, using (2.35), (2.33) and (2.37), one gets that

(A.5) $$\partial_N g_{\widetilde{\mathbb{S}}(Y,R)}(x) \geq \text{const}\, \frac{x_N - y_N}{|x-y|}.$$

Also, if $x$ is in the domain of $g_{\widetilde{\mathbb{S}}(Y,R)}$, then
$$|x-y| \leq \text{const}\, R.$$

The latter inequality, together with (A.5) and (A.3), ends the proof of (A.4).

Let now
$$\mathcal{R}(x_1, \ldots, x_{N-1}, x_{N+1}) :=$$
$$:= \left[ \left( H_0(x_{N+1}) - \frac{\overline{C_0}}{2R}(x_{N+1} - y_{N+1})^2 + R - H_0(y_{N+1}) \right)^2 - \sum_{j=1}^{N-1} |x_i - y_i|^2 \right]^{1/2}.$$

Let also $\pi_N : \mathbb{R}^{N+1} \longrightarrow \{x_N = 0\}$ be the natural projection, i.e.,
$$\pi_N(x_1, \ldots, x_{N+1}) := (x_1, \ldots, x_{N-1}, 0, x_{N+1}).$$

We now show that

(A.6) $$\pi_N \Big|_{\widetilde{\mathbb{S}}(Y,R) \cap B_{3a}(X_0)} \text{ is a diffeomorphism.}$$

For proving this, take any
$$X = \left( x, g_{\widetilde{\mathbb{S}}(Y,R)}(x) \right) \in \widetilde{\mathbb{S}}(Y,R) \cap B_{3a}(X_0),$$
and consider
$$\pi_N(X) = \left( x_1, \ldots, x_{N-1}, 0, g_{\widetilde{\mathbb{S}}(Y,R)}(x) \right).$$
Then, by (2.31) and (2.33),
$$\pi_N^{-1}\left( x_1, \ldots, x_{N-1}, 0, g_{\widetilde{\mathbb{S}}(Y,R)}(x) \right) = X,$$
with $x_N$ so that
$$|x_N - y_N| = \mathcal{R}(x_1, \ldots, x_{N-1}, x_{N+1}),$$
where $\mathcal{R}$ has been defined here above. From (A.3), we get that $\mathcal{R} \geq \text{const}\, R > 0$, thus $\mathcal{R}$ is smooth for $X \in B_{3a}(X_0)$. Also, using again (A.3), we see that $x_N > y_N$ for any $X \in \widetilde{\mathbb{S}}(Y,R) \cap B_{3a}(X_0)$: that is
$$\pi_N^{-1}\left( x_1, \ldots, x_{N-1}, 0, g_{\widetilde{\mathbb{S}}(Y,R)}(x) \right) =$$
$$= \left( x_1, \ldots, x_{N-1}, y_N + \mathcal{R}(x_1, \ldots, x_{N-1}, x_{N+1}), x_{N+1} \right),$$
thence (A.6) is proved.

Let now $\pi_{N+1} : \mathbb{R}^{N+1} \longrightarrow \{x_{N+1} = 0\}$ be the natural projection, i.e.,
$$\pi_{N+1}(x_1, \ldots, x_{N+1}) := (x_1, \ldots, x_N, 0).$$

Notice that, if $X = (x, g_{\widetilde{\mathbb{S}}(Y,R)}(x)) \in \widetilde{\mathbb{S}}(Y,R)$, then $\pi_{N+1}(X) = (x,0)$ and so
$$\pi_{N+1}^{-1}(x,0) = \left(x, g_{\widetilde{\mathbb{S}}(Y,R)}(x)\right),$$
showing that

(A.7) $\qquad \pi_{N+1}\Big|_{\widetilde{\mathbb{S}}(Y,R) \cap B_{3a}(X_0)}$ is a diffeomorphism.

Thus, we define
$$T := \pi_{N+1}\Big|_{\widetilde{\mathbb{S}}(Y,R) \cap B_{3a}(X_0)} \circ \pi_N^{-1}\Big|_{\widetilde{\mathbb{S}}(Y,R) \cap B_{3a}(X_0)}$$

Let also introduce the following domains:
$$O_1 := T\Big(\{x_N = 0\} \cap \{|x_{N+1}| < 3/4\} \cap B_{a+2}(\pi_N(X_0))\Big)$$
$$O_2 := T\Big(\{x_N = 0\} \cap \{|x_{N+1}| < 5/8\} \cap B_{a+1}(\pi_N(X_0))\Big).$$

Of course, $x_0 \in O_2 \subset O_1$ and, more precisely, by (A.6) and (A.7),

(A.8) $\qquad\qquad\qquad \mathrm{dist}\,(O_2, \partial O_1) \geq \mathrm{const}.$

Let us now notice that, in the light of Proposition 2.18,
$$\Delta_p g_{\widetilde{\mathbb{S}}(Y,R)} - \Delta_p u \leq \left(h_0'(g_{\widetilde{\mathbb{S}}(Y,R)}) - h_0'(u)\right) + \frac{\mathrm{const}}{R} \leq$$
$$\leq \Lambda\left((g_{\widetilde{\mathbb{S}}(Y,R)} + \frac{1}{R}) - u\right).$$

Hence, if $X \in O_1$,
$$-\Delta_p u + \Lambda u \leq -\Delta_p\left(g_{\widetilde{\mathbb{S}}(Y,R)} + \frac{1}{R}\right) + \Lambda\left(g_{\widetilde{\mathbb{S}}(Y,R)} + \frac{1}{R}\right),$$
where $\Lambda := \sup_{[-3/4,\, 3/4]} |h_0'|$. Hence, from the Harnack-type comparison inequality for[1] $p$-Laplacian (see, for instance [9], [35] or [10]), we get that

(A.9) $\qquad\qquad\qquad \sup_{O_2}(g_{\widetilde{\mathbb{S}}(Y,R)} - u) \leq \frac{C'}{R},$

for a suitable $C' > 1$, which may also depend on $a$.

Fix now $Z \in L \cap B_a(\pi_N X_0)$. Then, from (A.6), there exists
$$X^{(1)} = (x^{(1)}, x^{(1)}_{N+1}) \in \widetilde{\mathbb{S}}(Y,R) \cap B_{3a}(X_0)$$
so that
$$\pi_N\Big|_{\widetilde{\mathbb{S}}(Y,R) \cap B_{3a}(X_0)}(X^{(1)}) = Z,$$
that is

(A.10) $\qquad\qquad\qquad X^{(1)} = Z + t^{(1)} e_N$

for some $t^{(1)} \in \mathbb{R}$, and

(A.11) $\qquad\qquad\qquad X^{(1)} \in \widetilde{\mathbb{S}}(Y,R).$

---

[1] We recall that the gradient of $u$ does not vanish in the region we are considering, so that the assumptions needed in [9, 35, 10] are fulfilled.

Also, by (A.6),

$$|X^{(1)} - X_0| \leq \left|\left(\pi_N\big|_{\widetilde{\mathbb{S}}(Y,R) \cap B_{3a}(X_0)}\right)^{-1} (Z - \pi_N X_0)\right| \leq$$
(A.12)
$$\leq \text{const} |Z - \pi_N X_0| \leq$$
$$\leq \text{const} \, a \, .$$

Moreover, from (A.4), we have that, for any $t \geq 0$,

$$\begin{aligned} g_{\widetilde{\mathbb{S}}(Y,R)}(x^{(1)} + t e_N) &\geq \text{const} \, t + g_{\widetilde{\mathbb{S}}(Y,R)}(x^{(1)}) = \\ &= \text{const} \, t + x^{(1)}_{N+1} = \\ &= \text{const} \, t + z_{N+1} \, . \end{aligned}$$

Therefore, from (A.9),

(A.13) $$u(x^{(1)} + t e_N) > z_{N+1} \, ,$$

provided that $t \geq C''/R$, for a suitable $C'' > 1$, which may also depend on $a$. Analogously,

(A.14) $$u(x^{(1)} - t e_N) < z_{N+1} \, ,$$

provided that $t \geq C''/R$. From (A.13) and (A.14), we deduce the existence of $t^{(2)} \in [-C''/R, C''/R]$ so that

$$u(x^{(1)} + t^{(2)} e_N) = z_{N+1} \, .$$

Let us define $X^{(2)} := X^{(1)} + t^{(2)} e_N$. The point $x^{(2)} = x^{(2)}(Z)$ will be the one satisfying the thesis of Lemma 4.1, as we are now going to show. Notice that, by construction,

(A.15) $$\left|X^{(1)} - X^{(2)}\right| \leq \frac{C''}{R} \quad \text{and}$$
$$x^{(2)}_{N+1} = z_{N+1} = u(x^{(2)}) \, .$$

In particular,

$$\pi_N\left(x^{(2)}, u(x^{(2)})\right) = Z$$

and $|x^{(2)} - x_0| \leq$

(A.16)
$$\leq |x^{(1)} - x_0| + \frac{C''}{R} \leq$$
$$\leq \text{const} \, a + \frac{C''}{R} \leq$$
$$\leq \text{const} \, a \, ,$$

thanks to (A.12).

We now show that

(A.17) $$\left(x^{(1)} - x_0\right) \cdot \frac{\nabla u(x_0)}{|\nabla u(x_0)|} \leq H_0(z_{N+1}) - H_0(u(x_0)) + \frac{C'''}{R} \, ,$$

for some $C''' > 0$ which may depend on $a$. To prove (A.17), let us define

$$w(x) := H_0(g_{\widetilde{\mathbb{S}}(Y,R)}(x))$$

and notice that, from (A.10) and (A.11),
$$z_{N+1} = x_{N+1}^{(1)} = g_{\widetilde{\mathbb{S}}(Y,R)}(x^{(1)})$$
and so, since $X_0$ is a point where the graph of $u$ and $\widetilde{\mathbb{S}}(Y,R)$ touches,
$$\begin{aligned} H_0(z_{N+1}) - H_0(u(x_0)) &= \\ &= H_0\big(g_{\widetilde{\mathbb{S}}(Y,R)}(x^{(1)})\big) - H_0\big(g_{\widetilde{\mathbb{S}}(Y,R)}(x_0)\big) = \\ &= w(x^{(1)}) - w(x_0) \geq \\ &\geq \nabla w(x_0) \cdot (x^{(1)} - x_0) - \text{const}\, |D^2 w(\xi)| |x^{(1)} - x_0|^2\,, \end{aligned}$$
for some $\xi$ lying on the segment joining $x^{(1)}$ and $x_0$. Notice now that, by the definition of $w$ and the fact that $X_0$ is a point of touching between the graph of $u$ and $\widetilde{\mathbb{S}}(Y,R)$,
$$\begin{aligned} \nabla w(x_0) \cdot (x^{(1)} - x_0) &= \\ &= H_0'\big(g_{\widetilde{\mathbb{S}}(Y,R)}(x_0)\big) \nabla g_{\widetilde{\mathbb{S}}(Y,R)}(x_0) \cdot (x^{(1)} - x_0) = \\ &= H_0'\big(g_{\widetilde{\mathbb{S}}(Y,R)}(x_0)\big) \big|\nabla g_{\widetilde{\mathbb{S}}(Y,R)}(x_0)\big| \frac{\nabla g_{\widetilde{\mathbb{S}}(Y,R)}(x_0)}{\big|\nabla g_{\widetilde{\mathbb{S}}(Y,R)}(x_0)\big|} \cdot (x^{(1)} - x_0) = \\ &= H_0'\big(g_{\widetilde{\mathbb{S}}(Y,R)}(x_0)\big) \big|\nabla g_{\widetilde{\mathbb{S}}(Y,R)}(x_0)\big| \frac{\nabla u(x_0)}{|\nabla u(x_0)|} \cdot (x^{(1)} - x_0)\,, \end{aligned}$$
hence, from (2.53) and the fact that $X^{(1)} \in B_{3a}(X_0)$,
$$\nabla w(x_0) \cdot (x^{(1)} - x_0) \geq \frac{\nabla u(x_0)}{|\nabla u(x_0)|} \cdot (x^{(1)} - x_0) - \frac{\text{const}}{R}\, a\,.$$
Also, a direct computation and (2.54) imply that
$$\begin{aligned} \partial_{ij} w &= H_0''\big(g_{\widetilde{\mathbb{S}}(Y,R)}\big) \partial_i g_{\widetilde{\mathbb{S}}(Y,R)} \partial_j g_{\widetilde{\mathbb{S}}(Y,R)} + \\ &\quad + H_0'\big(g_{\widetilde{\mathbb{S}}(Y,R)}\big) \partial_{ij} g_{\widetilde{\mathbb{S}}(Y,R)} \leq \\ &\leq \frac{\text{const}}{R}\,. \end{aligned}$$
Collecting the estimates above and recalling that $X^{(1)} \in B_{3a}(X_0)$, the claim in (A.17) now easily follows.

Now, by means of (A.15) and (A.17),
$$\begin{aligned} \big(x^{(2)} - x_0\big) \cdot \frac{\nabla u(x_0)}{|\nabla u(x_0)|} &\leq \\ \leq \big(x^{(1)} - x_0\big) \cdot \frac{\nabla u(x_0)}{|\nabla u(x_0)|} + \big|x^{(1)} - x^{(2)}\big| &\leq \\ \leq H_0(z_{N+1}) - H_0(u(x_0)) + \frac{C''''}{R} &= \\ = H_0(u(x^{(2)})) - H_0(u(x_0)) + \frac{C''''}{R}\,, \end{aligned}$$
for a suitable $C'''' > 1$, which may depend on $a$. This, together with (A.12) and (A.16), completes the proof of Lemma 4.1.

## A.2. Proof of Lemma 4.2

The proof of Lemma 4.2 relies on an auxiliary result, namely Lemma A.1 here below, which may be seen as a rotation of the desired claim (see below (A.20)) plus a Lipschitz property on level sets[2]. For stating Lemma A.1, we need to introduce the following notation. Given $R > 0$ and $Y = (y, y_{N+1}) \in \mathbb{R}^N \times [-1/4, 1/4]$, we define $\Sigma(Y, R)$ as the zero level set of $\mathbb{S}(Y, R)$, that is:

$$\Sigma(Y, R) := \mathbb{S}(Y, R) \cap \{x_{N+1} = 0\} = \{g_{\mathbb{S}(Y,R)} = 0\}.$$

By the definitions on page 14 and (2.39), we have that $\Sigma(Y, R)$ is an $(N-1)$-dimensional sphere, namely

$$\Sigma(Y, R) = \{x \in \mathbb{R}^N \mid |x - y| = r\}$$

with[3]

(A.18) $$r = r(Y, R) := R - H_0(y_{N+1}) - \frac{\overline{C}_0}{2R} y_{N+1}^2.$$

The study of the geometry of such spheres is indeed linked with the study of the level sets of $\mathbb{S}(Y, R)$, via the following observation. If $s = g_{\mathbb{S}(Y,R)}(x) = g_{\widetilde{\mathbb{S}}(Y,R)}(x) \in (-1/2, 1/2)$, then, by (2.31) and Definition 2.15,

$$H_0(y_{N+1}) + |x - y| - R = H_0(s) - \frac{\overline{C}_0}{2R}(s - y_{N+1})^2,$$

hence, if $|s| < 1/2$, the signed distance between the $s$-level set of $g_{\mathbb{S}(Y,R)}$ and $\Sigma(Y, R)$ is given by

(A.19) $$H_0(s) + \frac{\overline{C}_0 s}{2R}(2y_{N+1} - s).$$

Given $x \in \mathbb{R}^N$, we now define

$$T_{Y,R} x = T_{Y,R}(x)$$

as the intersection point between $\Sigma(Y, R)$ and the half-line from $y$ going through $x$. With this, we can now deal with the above mentioned auxiliary result:

---

[2] The very rough idea underneath the proof of Lemma 4.2 goes as follows. First, Lemma A.1 provides a result which looks like a rotation of Lemma 4.2, and which possesses a uniform Lipschitz graph property for level sets. The proof of Lemma 4.2 will then ended by rotating back to the configuration in Lemma A.1: the Lipschitz property will take into account the error done in such rotation.

The idea for proving Lemma A.1 is that we would like to replace the estimates on the touching point set $\Xi$ with estimates on a suitable *first occurrence* touching point set $\breve{\Xi}$, i.e., with a set obtained by translating in the $e_N$ direction an appropriate barrier until it touches the graph of $u$. This strategy will present two advantages. First, the first occurrence touching property will easily imply the Lipschitz property for level sets of $\breve{\Xi}$ (which, as mentioned above, is needed for deducing Lemma 4.2 from Lemma A.1). Second, measure estimates for $\breve{\Xi}$ can be directly deduced from Proposition 3.14. For performing this, however, a technical difficulty arises: indeed, in order to be able to replace $\Xi$ with $\breve{\Xi}$, a "tiny" improvement of the assumptions of Lemma A.1 will be needed, namely (A.29) here below. Unfortunately, the proof of this detail is non trivial, and it will take several pages.

[3] Obviously, $r \sim R$ if $R$ is large.

LEMMA A.1. *Let $\check{C} > 1$ be a suitably large constant. Let $u$ be a $C^1$-subsolution of (1.5) in $\{|x'| < l\} \times \{|x_N| < l\}$. Assume that $\mathbb{S}(Y, R)$ is above the graph of $u$ in $\{|x'| \leq l\} \times \{|x_N| \leq l/2\}$ and that $\mathbb{S}(Y, R)$ touches the graph of $u$ at the point $(x_0, u(x_0))$. Suppose that*

- $|u(x_0)| < 1/2$, $|x_{0N}| < l/4$, $|x_0'| < l/4$;
- $\angle\left(\dfrac{\nabla u(x_0)}{|\nabla u(x_0)|}, e_N\right) \leq \dfrac{\pi}{8}.$

*Assume also that*

(A.20)
$$T_{Y,R} x_0 \in \{|x'| = q\} \cap \{x_N = 0\} \quad \text{and}$$
$$y = -e_N \sqrt{r^2 - q^2} \quad \text{with}$$
$$r = r(Y, R) = R - H_0(y_{N+1}) - \frac{\overline{C}_0}{2R} y_{N+1}^2.$$

*Then, there exist universal constants $C_1$, $C_2 > 1 > c > 0$ such that, if*

(A.21)
$$C_1 \leq q \leq \frac{l}{C_1} \quad \text{and} \quad 4\sqrt[3]{R} \leq l \leq cR,$$

*the following holds. Let $\Xi$ be the set of points $(\mathfrak{x}, u(\mathfrak{x})) \in \mathbb{R}^N \times \mathbb{R}$ satisfying the following properties:*

- $|\mathfrak{x}'| < q/15$, $|u(\mathfrak{x})| < 1/2$, $|\mathfrak{x} - x_0| < \check{C}q$;
- *there exists $\hat{Y} \in \mathbb{R}^{N+1}$ such that $\mathbb{S}(\hat{Y}, R/C_2)$ is above $u$ and it touches $u$ at $(\mathfrak{x}, u(\mathfrak{x}))$;*
- $\angle\left(\dfrac{\nabla u(\mathfrak{x})}{|\nabla u(\mathfrak{x})|}, \dfrac{\nabla u(x_0)}{|\nabla u(x_0)|}\right) \leq \dfrac{C_1 q}{R};$
- $(\mathfrak{x} - x_0) \cdot \dfrac{\nabla u(x_0)}{|\nabla u(x_0)|} \leq \dfrac{C_1 q^2}{R} + H_0(u(\mathfrak{x})) - H_0(u(x_0)).$

*Then,*

(A.22)
$$\mathcal{L}^N\left(\pi_N(\Xi)\right) \geq c q^{N-1}.$$

*More precisely, for any $s \in (-1/2, 1/2)$, there exists a set $\Xi_s \subseteq \Xi \cap \{x_{N+1} = s\}$, which is contained in a Lipschitz graph in the $e_N$-direction, with Lipschitz constant less than 1, and so that, if*

$$\check{\Xi} := \bigcup_{s \in (-1/2, 1/2)} \Xi_s,$$

*we have*[4]

(A.23)
$$\mathcal{L}^N\left(\pi_N(\check{\Xi})\right) \geq c q^{N-1}.$$

---

[4] Of course, (A.23) and the fact that $\check{\Xi} \subseteq \Xi$ imply (A.22).

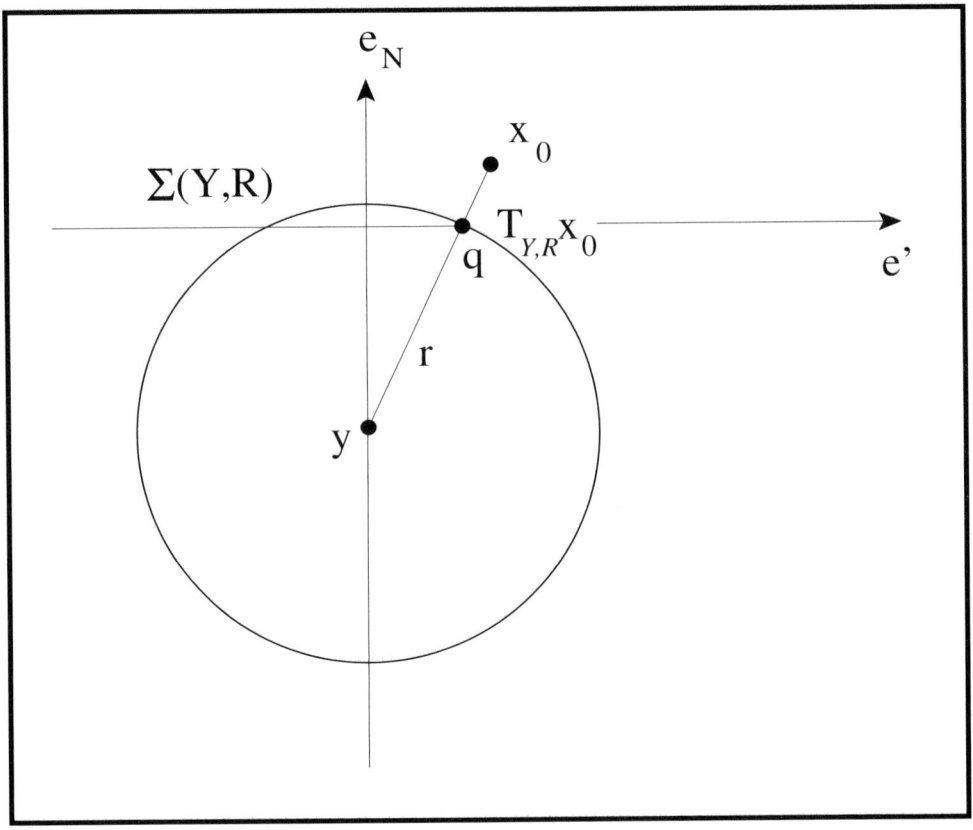

**The geometry of (A.20)**

PROOF. For further reference, let us point out some geometric features linking $x_0$ with its projection $T_{Y,R}x_0$. First of all, by (A.19),

(A.24) $$\left|T_{Y,R}x_0 - x_0\right| \leq \text{const}.$$

Also, by construction,

(A.25) $$\sin\left(\angle(x_0 - y, e_N)\right) = \frac{q}{r}.$$

Thus, (A.24) and (A.25) yield that

(A.26) $$\begin{aligned}|x_0'| &\leq \left|(T_{Y,R}x_0 - x_0)'\right| + \left|(T_{Y,R}x_0)'\right| \leq \\ &\leq \left|T_{Y,R}x_0 - x_0\right|\sin\left(\angle(x_0, e_N)\right) + q \leq \\ &\leq \frac{\text{const}\, q}{R} + q.\end{aligned}$$

Let us also observe that, if $|\mathfrak{x}'| \leq l/8$ and $\mathfrak{x}_N \in [-l/2, -l/4]$, then

$$|\mathfrak{x} - y| \leq r - \frac{l}{8} + \text{const} \leq R - \text{const}\sqrt[3]{R}$$

and thus, exploiting (2.15),

$$u(\mathfrak{x}) \leq g_{\mathbb{S}(Y,R)}(\mathfrak{x}) \leq$$
(A.27)
$$\leq g_{y_{N+1},R}\left(\operatorname{const}\left(1 - \sqrt[3]{R}\right)\right) =$$
$$= s_R.$$

More precisely, we have that

(A.28) $$u(\mathfrak{x}) < s_R$$

for any $\mathfrak{x}$ so that $|\mathfrak{x}'| \leq l/8$ and $\mathfrak{x}_N \in [-l/2, -l/4]$: indeed, if not, by (A.27) there would be a point for which the graph of $u$ touches the $s_R$-level from below and then a contradiction follows by applying Theorem B.6 to the function $s_R - u$ (see, e.g., the argument on page 18).

We now fix $C_* > 1$, to be chosen conveniently large. The first step of the proof of Lemma A.1 consists in proving the existence of a suitable $Y_* \in \mathbb{R}^{N+1}$ and $R_* > R/\operatorname{const}$, so that $\pi_{e_N} Y_* = \pi_{e_N} Y$ and $\mathbb{S}(Y_*, R_*)$ touches the graph of $u$ from above, in the region $\{|x'| \leq \check{C}q\}$, at the point $(x_*, u(x_*))$, with $|x'_*| \leq \check{C}q$ and

(A.29) $$T_{Y_*,R_*} x_* \in \left\{x_N \leq \frac{\operatorname{const} q^2}{R}\right\} \times \left\{|x'| < \frac{q}{C_*}\right\}.$$

We will prove (A.29) by iteration. Namely, we will set $Y_0 := Y$, $R_0 := R$ and, for any $k \in \mathbb{N}$, we will inductively find $Y_{k+1} \in \mathbb{R}^{N+1}$ and $R_{k+1} > R_k/4$, so that $\pi_{e_N} Y_{k+1} = \pi_{e_N} Y_k$ and $\mathbb{S}(Y_{k+1}, R_{k+1})$ touches the graph of $u$ from above at the point $(x_{k+1}, u(x_{k+1}))$, with $|x'_{k+1}| \leq \check{C}q$ and

(A.30) $$T_{Y_{k+1},R_{k+1}} x_{k+1} \in \left\{x_N \leq \frac{\operatorname{const} q^2}{R_k}\right\} \times \left\{|x'| < \eta q\right\},$$

for some $\eta \in (0,1)$. Since (A.29) follows by iterating (A.30) a finite number of times, we focus now on the proof of (A.30). More precisely, we will proof the first step in (A.30), i.e., the step with $k = 0$, since the others are analogous. The proof of (A.30) is actually quite non trivial, and it will take several pages (it will be ended on page 111).

For proving (A.30), let us begin by noticing that, if $|s| < 1/2$, from Definition 2.5 and (2.1) we get that

$$|h'_{s_0,R}(s) - h'_0(s)| = |\varphi'_{s_0,R}(s) - h'_0(s)| \leq$$
$$\leq |h'_0(s)| \left(\frac{|R^p - (\star)^p|}{(\star)^p} + \frac{\overline{C}_0 |s - s_0| \left(p/(p-1)\right)^{1/p}}{\left(h_0(s)\right)^{(p-1)/p} (\star)^{p+1}}\right),$$

where used the short hand notation

$$\star := R - \overline{C}_0 (s - s_0) \left(\frac{p}{p-1} h_0(s)\right)^{1/p}.$$

Therefore,

(A.31) $$|h'_{s_0,R}(s) - h'_0(s)| \leq \frac{\operatorname{const}}{R}, \text{ for any } |s| < 1/2.$$

Moreover, from Definition 2.5,

$$(A.32) \quad h'_{s_0,R}(s) = h'_0(s) - \frac{\widehat{C}_0}{R}, \text{ for any } s \in (s_R, -1/2) \cup (1/2, 1).$$

and, by construction, $\widehat{C}_0$ may be taken large if so is $\overline{C}_0$. We now fix a small parameter $\epsilon \in (0, 1/2]$ and a large parameter $\gamma > 1$ and we define

$$(A.33) \quad \omega := 2^{-1/(\gamma+2)} \text{ and } \eta := 1 - \epsilon(1 - \omega).$$

By construction, $\omega \in (0, 1)$ and $\eta \in (1/2, 1)$. Also, $\eta > \omega$ and, for $\gamma$ large, $\omega$ and $\eta$ are close to 1. For any $t > 0$, set also

$$\tilde{\psi}(t) := \frac{1}{\gamma}\left(\frac{1}{t^\gamma} - 1\right)$$

and, for any $z' \in \mathbb{R}^{N-1} \setminus \{0\}$,

$$\psi(z') := \tilde{\psi}(|z'|).$$

We now consider the graph

$$\mathfrak{G} := \left\{X \in \mathbb{R}^N \mid x_N = \frac{q^2}{\sqrt{r^2 - q^2}} \psi(x'/q)\right\}.$$

Since $\tilde{\psi}$ is strictly concave, while $\Sigma(\cdot, \cdot)$ is strictly convex, one sees that $\mathfrak{G}$ touches $\Sigma(Y, R)$ from above when $|x'| = q$. Analogously, if

$$y_\omega := -\left[\frac{q^2}{\gamma\sqrt{r^2 - q^2}}\left(1 - \frac{1}{\omega^\gamma}\right) + \omega^{\gamma+2}\sqrt{r^2 - q^2}\right] e_N \quad \text{and}$$

$$r_\omega := \omega^{\gamma+2}\sqrt{r^2 + q^2(\omega^{-2\gamma-2} - 1)},$$

one sees that $\mathfrak{G}$ touches the $(N-1)$-dimensional sphere $\partial B_{r_\omega}(y_\omega)$ from above when $|x'| = \omega q$. Notice that, by construction,

$$(A.34) \quad r_\omega \geq \omega^{\gamma+2} r = r/2,$$

and, more precisely, $r_\omega \in [r/2, \omega r]$ and $y_{\omega N} \geq y_N$. Also, from the fact that

$$(A.35) \quad \left|\sqrt{1+\tau} - 1 - \frac{\tau}{2}\right| \leq \tau^2,$$

provided that $\tau \in \mathbb{R}$ with $|\tau|$ sufficiently small, one sees that $y_N + r < y_{\omega N} + r_\omega$, if $r/q$ and $\gamma$ are suitably large. Notice also that, by construction,

$$(A.36) \quad \begin{aligned} |y_\omega - y| &= \frac{1}{2}\sqrt{r^2 - q^2} + \frac{q^2}{\gamma\sqrt{r^2 - q^2}}\left(\frac{1}{\omega^\gamma} - 1\right) \in \\ &\in \left[\frac{1}{2}\sqrt{r^2 - q^2}, \frac{1}{2}\sqrt{r^2 - q^2} + \frac{2q^2}{\gamma r}\right], \end{aligned}$$

if $\gamma$ and $r/q$ are large enough. Thus, we now consider the surface

$$\Gamma := \Gamma_1 \cup \Gamma_2 \cup \Gamma_3$$

defined in this way: we take

$$\begin{aligned} \Gamma_1 &:= \Sigma(Y, R) \cap \{x_N < 0\} \\ \Gamma_2 &:= \mathfrak{G} \cap \{|x'| \in [\omega q, q]\} \quad \text{and} \\ \Gamma_3 &:= \partial B_{r_\omega}(y_\omega) \cap \{|x'| < \omega q\} \cap \{x_N > 0\}. \end{aligned}$$

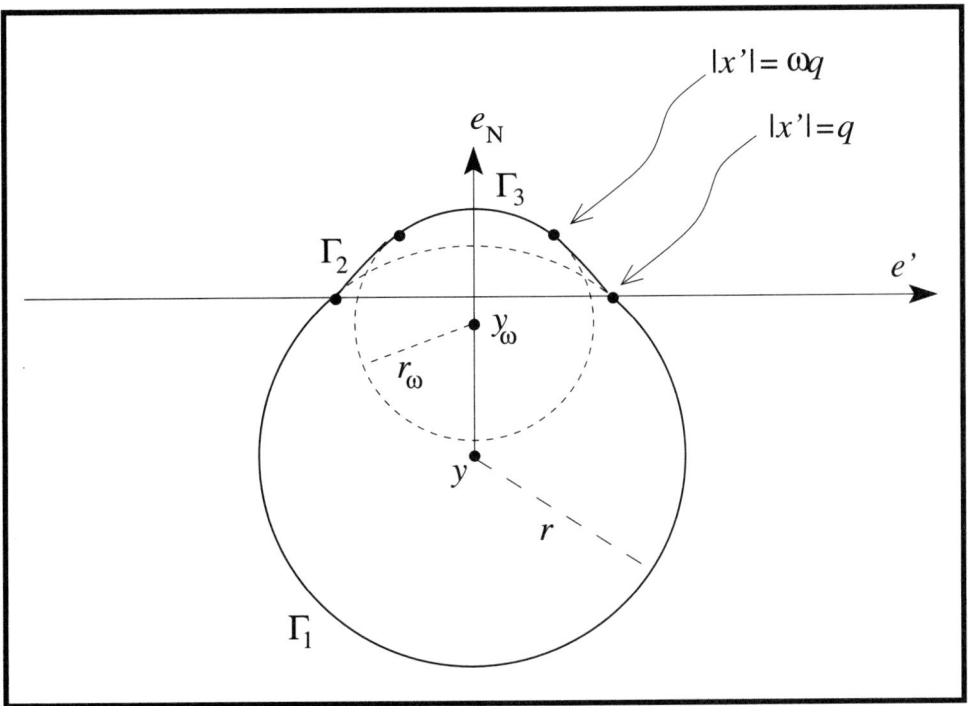

**The Barbapapa-like surface $\Gamma$**

In the sequel, we will often speak about points "inside (or outside) $\Gamma$", with the obvious meaning of points "inside (or outside) the bounded region whose boundary is $\Gamma$".

Also, by the above mentioned touching properties between $\mathfrak{G}$, $\Sigma(Y,R)$ and $\partial B_{r_\omega}(y_\omega)$, we have that $\Gamma$ is a $C^{1,1}$ closed hypersurface in $\mathbb{R}^N$. We denote by $d_\Gamma$ the signed distance to $\Gamma$, with the convention that $d_\Gamma$ is positive in the exterior of $\Gamma$ and negative in the interior.

The rough idea for proving (A.30) consists now in trying to find contact points whose projection on the zero level set of the corresponding barrier lies close[5] to $\Gamma_3$, and then use the geometry $\Gamma_3$, which is quite transparent.

We now define the following hypersurface in $\mathbb{R}^{N+1}$:

$$\Psi := \left\{ X \in \mathbb{R}^{N+1} \mid x_{N+1} = g_{y_{N+1},R}\Big(d_\Gamma(x) + H_{y_{N+1},R}(0)\Big) \right\}.$$

For the sake of simplicity, we will set

$$g_\Psi(x) := g_{y_{N+1},R}\Big(d_\Gamma(x) + H_{y_{N+1},R}(0)\Big),$$

---

[5]This is the reason also for introducing $\Sigma(Y_\omega, R_1)$ later on (see page 98).

so[6] that $\Psi = \{x_{N+1} = g_\Psi(x)\}$. Note that
$$d_\Gamma(x) \leq |x - y| - r,$$
thus, from (2.39) and (2.31),

(A.37) $$g_\Psi \leq g_{\mathbb{S}(Y,R)}.$$

Let us now show some further properties of $\Psi$. First of all, $\Psi$ coincides with $\mathbb{S}(Y, R)$ at any points for which $d_\Gamma$ is realized on $\Gamma_1$; more precisely,

(A.38) $$\text{if } d_\Gamma(x) = d_{\Gamma_1}(x), \text{ then } g_\Psi(x) = g_{\mathbb{S}(Y,R)}(x).$$

Indeed, if $d_\Gamma(x) = d_{\Gamma_1}(x)$, by some geometric considerations and (2.39), we have that
$$\begin{aligned}
g_\Psi(x) &= g_{y_{N+1},R}\Big(d_{\Gamma_1}(x) + H_{y_{N+1},R}(0)\Big) = \\
&= g_{y_{N+1},R}\Big(d_{\Gamma_1}(x) - \frac{\overline{C}_0}{2R} y_{N+1}^2\Big) = \\
&= g_{y_{N+1},R}\Big(|x - y| - r - \frac{\overline{C}_0}{2R} y_{N+1}^2\Big).
\end{aligned}$$

This, (A.18) and (2.20) end the proof of (A.38).

In the light of (A.38), (A.37) and Proposition 2.13, we deduce that

(A.39)
$$\begin{aligned}
&\Delta_p g_\Psi(x) < h'_0(g_\Psi(x)) \text{ in the viscosity sense} \\
&\text{at any } x \in \mathbb{R}^N \text{ for which} \\
&d_\Gamma(x) \text{ is attained on } \Gamma_1 \text{ and} \\
&g_\Psi(x) \in [s_R, -1/2] \cap [1/2, 1).
\end{aligned}$$

Furthermore, in an appropriate system of coordinates (see §14.6 of [19] for details on the distance function),
$$\nabla g_\Psi(x) = g'_{y_{N+1},R}(\star) e_N$$
and $D^2 g_\Psi(x)$ is the $N \times N$ diagonal matrix with the following entries on the diagonal:
$$\frac{\kappa_1}{\kappa_1 d_\Gamma(x) - 1} g'_{y_{N+1},R}(\star), \ldots, \frac{\kappa_{N-1}}{\kappa_{N-1} d_\Gamma(x) - 1} g'_{y_{N+1},R}(\star), g''_{y_{N+1},R}(\star).$$

Here above, we denoted
$$\star := d_\Gamma(x) + H_{y_{N+1},R}(0),$$

---

[6] To have some further geometric insight, one may observe that the domain where $g_\Psi$ is defined and non-constant is in a $O(\log R)$-neighborhood of $\Gamma$ (recall (2.39), (2.31) and Lemma 2.10). The principal curvatures of $\Gamma$ are of order $R$. This implies that if $x$ is in the domain where $g_\Psi$ is defined and non-constant, then the distance from $x$ to $\Gamma$ is realized at exactly one point. Furthermore, since $R$ is much bigger than $l$ (recall (A.21)), then it may be convenient to look at $\Gamma \cap [-l, l]^N$ (that is, $\Gamma$ in the domain we are interested in) as a graph in the $e_N$-direction. By direct inspection, the slope of such graph at any point $\mathfrak{x}$ is of order $|\mathfrak{x}'|/R$. In particular, the angle between $e_N$ and the normal of $\Gamma$ at $\mathfrak{x}$ is of order $|\mathfrak{x}'|/R$. The above slope bound and (A.21) also imply that $\Gamma$ is quite flat in $[-l, l]^N$, namely, its slope is bounded by an order $c$.

while, as standard, $\kappa_1, \ldots, \kappa_{N-1}$ represent the principal curvatures of $\Gamma$ at the point where $d_\Gamma(x)$ is realized. Hence, from Definitions 2.8 and 2.11 and Lemma B.3, we get that

$$\text{(A.40)} \qquad \nabla g_\Psi(x) = \left(\frac{p}{p-1} h_{y_{N+1},R}(\sharp)\right)^{1/p} e_N,$$

where we denoted $\sharp := g_{y_{N+1},R}(\star)$, and that $D^2 g_\psi(x)$ may be represented as the $N \times N$ diagonal matrix with the following entries on the diagonal:

$$\text{(A.41)} \qquad \begin{aligned} &\frac{\kappa_1}{\kappa_1 d_\Gamma(x) - 1} \left(\frac{p}{p-1} h_{y_{N+1},R}(\sharp)\right)^{1/p} \\ &\vdots \\ &\frac{\kappa_{N-1}}{\kappa_{N-1} d_\Gamma(x) - 1} \left(\frac{p}{p-1} h_{y_{N+1},R}(\sharp)\right)^{1/p} \\ &\frac{(p h_{y_{N+1},R}(\sharp))^{(2-p)/p}}{(p-1)^{2/p}} h'_{y_{N+1},R}(\sharp). \end{aligned}$$

From (A.40) and (A.41),

$$\text{(A.42)} \qquad \begin{aligned} \Delta_p g_\Psi &= h'_{y_{N+1},R} + \\ &+ \left(\frac{p}{p-1} h_{y_{N+1},R}\right)^{\frac{p-1}{p}} \sum_{i=1}^{N-1} \frac{\kappa_i}{\kappa_i d_\Gamma - 1}, \end{aligned}$$

outside $\{|g_\Psi| = 1/2\} \cup \{\nabla g_\Psi = 0\}$, where we dropped the $\sharp$-dependence for the sake of simplicity. Let us now compute the principal curvatures $\kappa_i^{\Gamma_2}$ of the hypersurface $\Gamma_2$: exploiting Lemma B.14, we have that

$$\text{(A.43)} \qquad \begin{aligned} -\kappa_1^{\Gamma_2} = \cdots = -\kappa_{N-2}^{\Gamma_2} &= \frac{q^{\gamma+2}}{|x'|\sqrt{(r^2 - q^2)|x'|^{2\gamma+2} + q^{2\gamma+4}}} \geq 0 \\ \kappa_{N-1}^{\Gamma_2} &= \frac{(\gamma+1) q^{\gamma+2} (r^2 - q^2) |x'|^{2\gamma+1}}{\left((r^2 - q^2)|x'|^{2\gamma+2} + q^{2\gamma+4}\right)^{3/2}} \geq 0. \end{aligned}$$

In particular, since $|x'| \geq \omega q$ on $\Gamma_2$, we infer from the above relations that

$$\text{(A.44)} \qquad -\kappa_i^{\Gamma_2} \leq \frac{1}{r_\omega},$$

for $i = 1, \ldots, N-2$. Thus, we deduce from (A.34), (A.44) and (A.18) that

$$\text{(A.45)} \qquad -\kappa_i^{\Gamma_2} \leq \frac{3}{R},$$

for $i = 1, \ldots, N-2$, if $R$ is large enough. Furthermore, since $q \geq |x'| \geq \omega q$ on $\Gamma_2$, (A.43) gives

$$\begin{aligned} \kappa_{N-1}^{\Gamma_2} &\geq \frac{(\gamma+1)(r^2 - q^2) \omega^{2\gamma+1}}{r^3} \geq \\ &\geq \frac{9}{10} \frac{(\gamma+1) \omega^{2\gamma+1}}{r} \geq \\ &\geq \frac{9}{40} \frac{(\gamma+1)}{r}, \end{aligned}$$

that is

(A.46) $$\kappa_{N-1}^{\Gamma_2} \geq \text{const} \frac{(\gamma+1)}{R}.$$

Analogously, one sees that

(A.47) $$\kappa_{N-1}^{\Gamma_2} \leq \text{const} \frac{(\gamma+1)}{R}.$$

We now claim that

(A.48) $$\Delta_p g_\Psi(x) \leq h_0'(g_\Psi(x)) - \frac{\text{const } \gamma}{R}$$
at any $x \in \mathbb{R}^N$ for which $|g_\Psi(x)| \neq 1/2$,
$\nabla g_\Psi(x) \neq 0$,
$|d_\Gamma(x)| \leq 2\sqrt{R}$ and $d_\Gamma$ is attained on $\Gamma_2$.

To prove this, take $x$ as requested here above: then, thanks to (A.45), (A.47) and the fact that $|d_\Gamma(x)| \leq 2\sqrt{R}$, that

$$\left|\kappa_{N-1}^{\Gamma_2} d_\Gamma(x)\right| + \sum_{i=1}^{N-2} \left|\kappa_i^{\Gamma_2} d_\Gamma(x)\right| \leq \frac{\text{const}}{\sqrt{R}}$$

and so, if $R$ is large enough,

(A.49) $$\sum_{i=1}^{N-1} \frac{-\kappa_i^{\Gamma_2}}{1 - \kappa_i^{\Gamma_2} d_\Gamma(x)} \leq -2 \sum_{i=1}^{N-2} \kappa_i^{\Gamma_2} - \frac{\kappa_{N-1}^{\Gamma_2}}{2}.$$

Hence, using the regularity of the functions involved in our domain, (A.42), (A.49), (A.31) and (A.32),

$$\begin{aligned}
\Delta_p g_\Psi &= h'_{y_{N+1},R} + \left(\frac{p}{p-1} h_{y_{N+1},R}\right)^{\frac{p-1}{p}} \sum_{i=1}^{N-1} \frac{\kappa_i^{\Gamma_2}}{\kappa_i^{\Gamma_2} d_\Gamma - 1} \leq \\
&\leq h'_{y_{N+1},R} + \left(\frac{p}{p-1} h_{y_{N+1},R}\right)^{\frac{p-1}{p}} \left(-2 \sum_{i=1}^{N-2} \kappa_i^{\Gamma_2} - \frac{\kappa_{N-1}^{\Gamma_2}}{2}\right) \leq \\
&\leq h_0' + \frac{\text{const}}{R} + \\
&\quad + \left(\frac{p}{p-1} h_{y_{N+1},R}\right)^{\frac{p-1}{p}} \left(-2 \sum_{i=1}^{N-2} \kappa_i^{\Gamma_2} - \frac{\kappa_{N-1}^{\Gamma_2}}{2}\right).
\end{aligned}$$

We thus deduce, by (A.45) and (A.46), that

$$\begin{aligned}
\Delta_p g_\Psi &\leq h_0' + \frac{\text{const}}{R} + \\
&\quad + \left(\frac{p}{p-1} h_{y_{N+1},R}\right)^{\frac{p-1}{p}} \left(\frac{6(N-2)}{R} - \frac{\text{const}(\gamma+1)}{2R}\right).
\end{aligned}$$

Thus, if $\gamma$ is large enough,

$$\Delta_p g_\Psi \leq h_0' + \frac{\text{const}}{R} - \text{const} \left(\frac{p}{p-1} h_{y_{N+1},R}\right)^{\frac{p-1}{p}} \frac{(\gamma+1)}{4R}.$$

Therefore, since we are evaluating $h_{y_{N+1},R}$ at $g_\Psi(x) \in [-1/2, 1/2]$,

$$\Delta_p g_\Psi \leq h'_0 + \frac{\text{const}}{R} - \frac{\text{const}\,(\gamma+1)}{R},$$

which proves (A.48), if $\gamma$ is chosen to be conveniently large.

We now show that

(A.50) $\quad\quad\quad\Delta_p g_\Psi(x) < h'_0(g_\Psi(x))$ in the viscosity sense

at any $x \in \mathbb{R}^N$ for which

$d_\Gamma$ is attained on $\Gamma_2$.

For proving this, we first point out that we may assume

(A.51) $$g_\Psi(x) > s_R$$

Indeed, if $g_\Psi(x) = s_R$, arguing as in Proposition 2.13

$$\Delta_p g_\Psi(x) = 0 < h'_0(s_R) = h'_0(g_\Psi(x)),$$

in the viscosity sense, giving the desired claim.

For proving (A.50), we may also assume $\nabla g_\Psi(x) \neq 0$, otherwise, we would have $g_\Psi(x) = s_R$ and we go back to (A.51). We may also assume that $|d_\Gamma(x)| \leq \sqrt{R}$: indeed, if $|d_\Gamma(x)| \geq \sqrt{R}$, we have that

$$|d_\Gamma(x) + H_{y_{N+1},R}(0)| \geq \sqrt{R}/2 > C_1 \log R$$

and so, by Lemma 2.10, either

$$d_\Gamma(x) + H_{y_{N+1},R}(0) > H_{y_{N+1},R}(1),$$

in which case, due to Definition 2.11, $g_\Psi(x)$ is not even defined, or

$$d_\Gamma(x) + H_{y_{N+1},R}(0) < H_{y_{N+1},R}(s_R),$$

in which case $g_\Psi$ is constantly equal to $s_R$ in a neighborhood of $x$, which has just been ruled out. Also, in the proof of (A.50), we can restrict ourselves to the case in which $|g_\Psi| \neq 1/2$, since, by Proposition 2.13, no smooth function can touch $g_\Psi$ from below at level $\pm 1/2$. With these further (non restrictive) assumptions, it is easy to deduce (A.50) from (A.48).

We now prove that

(A.52) $\quad\quad\quad\Delta_p g_\Psi(x) < h'_0(g_\Psi(x))$ in the viscosity sense

at any $x \in \mathbb{R}^N$ for which $|g_\Psi(x)| \geq 1/2$ and

$d_\Gamma(x)$ is attained on $\Gamma_3$.

To prove the above claim, notice that, as remarked here above, we may restrict ourselves to the case in which $|g_\Psi| \neq 1/2$ and $|d_\Gamma(x)| \leq \sqrt{R}$. Also, since $\Gamma_3$ is a portion of sphere,

$$\kappa_1^{\Gamma_3} = \cdots = \kappa_{N-1}^{\Gamma_3} = -\frac{1}{r_\omega} < 0,$$

so, since $|d_\Gamma(x)| \leq \sqrt{R}$, we get that, for $i = 1, \ldots, N-1$,

$$\left|\kappa_i^{\Gamma_3} d_\Gamma\right| \leq \frac{\text{const}}{\sqrt{R}}$$

which is small; therefore, from (A.32),

$$\Delta_p g_\Psi =$$

$$= h'_{y_{N+1},R} + \left(\frac{p}{p-1} h_{y_{N+1},R}\right)^{\frac{p-1}{p}} \sum_{i=1}^{N-1} \frac{\kappa_i^{\Gamma_3}}{\kappa_i^{\Gamma_3} d_\Gamma - 1} \leq$$

$$\leq h'_{y_{N+1},R} + \text{const} \sum_{i=1}^{N-1} |\kappa_i^{\Gamma_3}| \leq$$

$$\leq h'_{y_{N+1},R} + \frac{\text{const}}{R} \leq$$

$$\leq h'_0 - \frac{\widehat{C}_0}{R} + \frac{\text{const}}{R} <$$

$$< h'_0,$$

provided that $\widehat{C}_0$ is chosen suitably large, thus proving (A.52).

In the light of (A.39), (A.50) and (A.52), we deduce that

(A.53) $\quad g_\Psi$ is a strict supersolution of (1.5) everywhere
possibly except the set $\{|g_\Psi| < 1/2\} \cap \left\{ d_\Gamma \text{ realized on } \Gamma_1 \cup \Gamma_3 \right\}.$

We now point out that

(A.54) $\qquad\qquad$ if $|x'| \geq \check{C}q$, then $g_\Psi(x) = g_{\mathbb{S}(Y,R)}(x).$

Indeed, let $\mathfrak{x} \in \Gamma$ realize $d_\Gamma(x)$. Then, if $\nu$ is the outer normal of $\Gamma$ at $\mathfrak{x}$, we have that $\angle(\nu, e_N) \leq \text{const } |\mathfrak{x}'|/R$ (see the footnote on page 94). Therefore,

$$|\mathfrak{x}' - x'| = |\mathfrak{x} - x| \sin\left(\angle(\nu, e_N)\right) \leq \frac{\text{const } |\mathfrak{x}'| \, |\mathfrak{x} - x|}{R} \leq \frac{\text{const } |\mathfrak{x}'| \, l}{R} \leq \text{const } |\mathfrak{x}'|.$$

Since

$$|x' - \mathfrak{x}'| \geq \check{C}q - |\mathfrak{x}'|,$$

we thence deduce that

$$\check{C}q \leq \text{const } |\mathfrak{x}'|,$$

thus $|\mathfrak{x}'| \geq 2q$. In particular, $\mathfrak{x} \in \Gamma_1$, therefore (A.54) follows from (A.38).

We now define

$$Y_\omega := (y_\omega, y_{N+1}),$$
$$R_1 := r_\omega + H_0(y_{N+1}) + \frac{5\overline{C}_0}{R},$$
$$r_1 := R_1 - H_0(y_{N+1}) - \frac{\overline{C}_0}{2R_1} y_{N+1}^2,$$
$$\Sigma(Y_\omega, R_1) := \mathbb{S}(Y_\omega, R_1) \cap \{x_{N+1} = 0\}.$$

By definition,

(A.55) $\qquad\qquad\qquad R_1 \leq \omega r + \text{const} \leq$
$\qquad\qquad\qquad\qquad \leq \omega R + \text{const} < R$

and

(A.56) $$r_1 - r_\omega \in \left[\frac{3\overline{C}_0}{R}, \frac{5\overline{C}_0}{R}\right].$$

Furthermore, arguing as done on page 88, we have that

(A.57) $$\Sigma(Y_\omega, R_1) = \partial B_{r_1}(y_\omega);$$

thus, by (A.56), we infer that

(A.58) $\Sigma(Y_\omega, R_1)$ stays at distance greater than $3\overline{C}_0/R$ outside $\partial B_{r_\omega}(y_\omega) \supset \Gamma_3$.

Recalling the definition of $\eta$ given in (A.33), we now show that

(A.59) $\partial B_{r_\omega}(y_\omega) \cap \{|x'| \geq \eta q\}$ is at distance at least $\dfrac{\text{const } q^2}{R}$ inside $\Gamma$.

For the proof of this, it is convenient, to think $\Gamma$ and $\partial B_{r_\omega}(y_\omega)$ (in $[-l, l]^N$) as graphs in the $e_N$-direction (see the footnote on page 94): we then explicitly compute the "vertical distance" from $\partial B_{r_\omega}(y_\omega)$ to $\Gamma$ (with the sign convention that such vertical distance is positive at points where $\partial B_{r_\omega}(y_\omega)$ is below $\Gamma$ in the $e_N$-direction) and compare it to the "true" distance by using the flatness of these graphs (in $[-l, l]^N$). To formalize such idea, we proceed as follows. We write $\Gamma$ and $\partial B_{r_\omega}(y_\omega)$ (in $[-l, l]^N$) as graphs in the direction $e_N$, that is, we consider $G_1, G_2 \in C^{1,1}([0, l])$ so that

$$\Gamma \cap [-l, l]^N = \left\{x_N = G_1(|x'|)\right\}$$

and

$$\partial B_{r_\omega}(y_\omega) \cap [-l, l]^N = \left\{x_N = G_2(|x'|)\right\}.$$

Then, we define the vertical distance between $\Gamma$ and $\partial B_{r_\omega}(y_\omega)$ as

$$G_1(|x'|) - G_2(|x'|).$$

To evaluate it, note that, by construction, $G_1(t) = G_2(t)$ if $t \in [0, \omega q]$ and $G_1'(\omega q) = G_2'(\omega q)$. What is more, since $G_2(t) = y_{\omega, N} + \sqrt{r_\omega^2 - t^2}$, one has that

$$G_2''(t) \leq -\frac{1 - \text{const } c}{r_\omega} \leq -\frac{4}{3} \cdot \frac{1 - \text{const } c}{r}.$$

Analogously, since $G_1(t) = y_N + \sqrt{r - t^2}$ for any $t \geq q$, one has that

$$G_1''(t) \geq -\frac{1 + \text{const } c}{r},$$

for any $t \geq q$. Also, by construction,

$$G_1''(t) \geq 0$$

for $t \in [\omega q, q]$. Let us define $G := G_1 - G_2$. By means of the above computations, we have that $G(\omega q) = G'(\omega q) = 0$, that $G''(t) \geq \text{const}/r$ for $t \in [\omega q, \eta q]$ and

$G''(t) \geq 0$ for $t \geq \eta q$. Therefore, if $t \geq \eta q$, then

$$G(t) = \int_{\omega q}^{t} \int_{\omega q}^{s} G''(\zeta) \, d\zeta \, ds =$$

$$= \int_{\omega q}^{t} (t - \zeta) G''(\zeta) \, d\zeta \geq$$

$$\geq \text{const} \int_{\omega q}^{\eta q} \frac{(\eta q - \zeta)}{r} \geq \frac{\text{const } q^2}{R}.$$

This says that, if $x \in \partial B_{r_\omega}(y_\omega)$ with $|x'| \geq \eta q$, then $x$ is inside $\Gamma$, with vertical distance greater than $\text{const } q^2 / R$. Thence, if $x \in \partial B_{r_\omega}(y_\omega)$ with $|x'| \geq \eta q$ and $z$ realizes $d_\Gamma(x)$, denoting by $w$ the point in $\Gamma$ so that $w' = x'$, we have that

(A.60) $$-d_\Gamma(x) = |z - x|$$

and that

(A.61) $$|x - w| \geq \frac{\text{const } q^2}{R}.$$

Elementary trigonometry and the flatness of $\Gamma$ and $B_{r_\omega}(y_\omega)$ in $[-l, l]^N$ (recall the footnote on page 94) also implies that

(A.62) $$|x - w| \leq 2|x - z|.$$

Then, (A.59) follows from (A.60), (A.61) and (A.62).

A first consequence of (A.59) is that, for any $a \in \partial B_{r_\omega}(y_\omega)$ and any $b \in \Gamma_1$,

(A.63) $$|a - b| \geq \text{const} \frac{q^2}{R}.$$

We also infer from (A.57), (A.59) and (A.56) that

(A.64) $$\Sigma(Y_\omega, R_1) \cap \{|x'| \geq \eta q\} \text{ is at distance at least } \frac{3\overline{C}_0}{R} \text{ inside } \Gamma.$$

Notice also that, by construction and recalling (A.35), one has

$$y_{\omega N} + r_1 \leq -\frac{1}{2}\sqrt{r^2 - q^2} + \frac{\text{const } q^2}{r} + r_\omega + \frac{\text{const } \overline{C}_0}{R} =$$

$$= \frac{1}{2}\left(\sqrt{r^2 + (4\omega^2 - 1)q^2} - \sqrt{r^2 - q^2}\right) +$$

$$+ \frac{\text{const } q^2}{r} + \frac{\text{const } \overline{C}_0}{R} \leq$$

$$\leq \frac{\text{const } q^2}{r} + \frac{\text{const } \overline{C}_0}{R} \leq$$

$$\leq \frac{\text{const } q^2}{R},$$

thence,

(A.65) $$\Sigma(Y_\omega, R_1) \subseteq \left\{x_N \leq \frac{\text{const } q^2}{R}\right\}.$$

## A.2. PROOF OF LEMMA 4.2

We now investigate the mutual position of $\Psi$ and $\mathbb{S}(Y_\omega, R_1)$. For this, we start by claiming that

(A.66)
$$\text{The region } \left\{ X \in \Psi \text{ where } |x_{N+1}| < 1/2 \text{ and } d_\Gamma(x) \text{ is realized on } \Gamma_3 \right\} \text{ is above } \mathbb{S}(Y_\omega, R_1)$$

(where above means with respect to the $e_{N+1}$-direction).

To prove (A.66), note that, if $d_\Gamma(x)$ is realized on $\Gamma_3$, then $x - y_\omega$ is orthogonal to $\Gamma_3 \subset \partial B_{r_\omega}(y_\omega)$, therefore, recalling (A.56) and (2.39),

$$\begin{aligned}
d_\Gamma(x) &= d_{\Gamma_3}(x) = d_{\Sigma(Y_\omega, R_1)}(x) + r_1 - r_\omega \geq \\
&\geq d_{\Sigma(Y_\omega, R_1)}(x) + \frac{3\overline{C}_0}{R} = \\
&= |x - y_\omega| - r_1 + \frac{3\overline{C}_0}{R} = \\
&= |x - y_\omega| - R_1 + H_0(y_{N+1}) + \frac{\overline{C}_0}{2R_1} y_{N+1}^2 + \frac{3\overline{C}_0}{R} = \\
&= |x - y_\omega| - R_1 + H_0(y_{N+1}) + \frac{\overline{C}_0}{2R_1} y_{N+1}^2 + \frac{3\overline{C}_0}{R} - \\
&\quad - H_{y_{N+1}, R}(0) - \frac{\overline{C}_0}{2R} y_{N+1}^2.
\end{aligned}$$

This, recalling (2.39) and (2.38), gives that

$$\begin{aligned}
&H_{y_{N+1}, R_1}\big(g_\Psi(x)\big) - \frac{\overline{C}_0}{2}\big(g_\Psi(x) - y_{N+1}\big)^2 \left(\frac{1}{R} - \frac{1}{R_1}\right) = \\
&= \widetilde{H}_{y_{N+1}, R_1}\big(g_\Psi(x)\big) - \frac{\overline{C}_0}{2}\big(g_\Psi(x) - y_{N+1}\big)^2 \left(\frac{1}{R} - \frac{1}{R_1}\right) = \\
&= \widetilde{H}_{y_{N+1}, R}\big(g_\Psi(x)\big) = \\
&= H_{y_{N+1}, R}\big(g_\Psi(x)\big) = \\
&= H_{y_{N+1}, R}\Big(g_{y_{N+1}, R}\big(d_\Gamma(x) + H_{y_{N+1}, R}(0)\big)\Big) = \\
&= d_\Gamma(x) + H_{y_{N+1}, R}(0) \geq \\
&\geq |x - y_\omega| - R_1 + H_0(y_{N+1}) + \frac{\overline{C}_0}{2R_1} y_{N+1}^2 + \frac{3\overline{C}_0}{R} - \frac{\overline{C}_0}{2R} y_{N+1}^2.
\end{aligned}$$

Therefore,

(A.67)
$$\begin{aligned}
H_{y_{N+1}, R_1}\big(g_\Psi(x)\big) &\geq |x - y_\omega| - R_1 + H_0(y_{N+1}) + \\
&\quad + \frac{3\overline{C}_0}{R} - \frac{\overline{C}_0}{2R} y_{N+1}^2 - \frac{\overline{C}_0}{2R_1}\big(g_\Psi(x) - y_{N+1}\big)^2.
\end{aligned}$$

Note however that $R_1 \sim R/2$ for large $R$, and so, in particular, we may assume that $R_1 \geq R/3$. Thence,

$$\frac{3\overline{C_0}}{R} - \frac{\overline{C_0}}{2R}y_{N+1}^2 - \frac{\overline{C_0}}{2R_1}\left(g_\Psi(x) - y_{N+1}\right)^2 \geq$$
$$\geq \frac{3\overline{C_0}}{R} - \frac{\overline{C_0}}{2R}\left(\frac{1}{4}\right)^2 - \frac{3\overline{C_0}}{2R}\left(\frac{3}{4}\right)^2 >$$
$$> 0.$$

This and (A.67) yield that

$$H_{y_{N+1}, R_1}\left(g_\Psi(x)\right) > |x - y_\omega| - R_1 + H_0(y_{N+1}),$$

and so

$$g_\Psi(x) > g_{y_{N+1}, R_1}\left(|x - y_\omega| - R_1 + H_0(y_{N+1})\right) = g_{\mathbb{S}(Y_\omega, R_1)}(x),$$

which proves (A.66).

We now show that

(A.68) The region $\Big\{X \in \mathbb{S}(Y_\omega, R_1)$ where $d_{\Sigma(Y_\omega, R_1)}(x)$ is realized at a point $z$ with $|z'| \geq \eta q\Big\}$ is strictly above $\Psi$ (in the $e_{N+1}$-direction).

In order to prove (A.68), take $X \in \mathbb{S}(Y_\omega, R_1)$ and assume that

$$\sigma := \text{dist}\left(x, \Sigma(Y_\omega, R_1)\right)$$

is attained at $z \in \Sigma(Y_\omega, R_1)$ and $|z'| \geq \eta q$. Then, by construction,

$$x_{N+1} = g_{\mathbb{S}(Y_\omega, R_1)}(x),$$

thus from (2.20) and Definition 2.11,

$$|x - y_\omega| = H_0(x_{N+1}) - \frac{\overline{C_0}}{2R_1}(x_{N+1} - y_{N+1})^2 + R_1 - H_0(y_{N+1}).$$

Therefore,

$$\text{dist}\left(x, \Sigma(Y_\omega, R_1)\right) = |x - y_\omega| - r_1 =$$
$$= H_0(x_{N+1}) - \frac{\overline{C_0}}{2R_1}(x_{N+1} - y_{N+1})^2 + R_1 - H_0(y_{N+1}) - r_1,$$

which implies that

$$\sigma \leq H_{y_{N+1}, R}(x_{N+1}) + \frac{\overline{C_0}}{2}\left(\frac{1}{R} + \frac{1}{R_1}\right).$$

Hence,

(A.69) $$g_{\mathbb{S}(Y_\omega, R_1)}(x) = x_{N+1} \geq g_{y_{N+1}, R}\left(\sigma - \frac{\overline{C_0}}{2}\left(\frac{1}{R} + \frac{1}{R_1}\right)\right).$$

## A.2. PROOF OF LEMMA 4.2

We consider $\check{y} \in \Gamma$ on the half-line from $y_\omega$ towards $z$. Note that, by construction, $y_\omega$, $z$, $x$ and $\check{y}$ lie on the same half-line. Let also $\tilde{y} \in \Gamma$ be the point realizing $d_\Gamma(x)$. By means of (A.64), we have that

(A.70) $$|z - \check{y}|, \ |z - \tilde{y}| \geq \frac{3\overline{C}_0}{R}.$$

We claim that this implies that

(A.71) $$\sigma \geq d_\Gamma(x) + \frac{3\overline{C}_0}{R}.$$

To prove this, we distingush three cases, according to the mutual position of $x$, $z$ and $\check{y}$. Namely, if $x$ is outside $\Gamma$ and outside $\Sigma(Y_\omega, R_1)$, we have that

$$\begin{aligned} \sigma - d_\Gamma(x) &= |x-z| - |d_\Gamma(x)| \geq \\ &\geq |x-z| - |x-\check{y}| = \\ &= |\check{y}-z|, \end{aligned}$$

and so (A.71) follows from (A.70) in this case; if, on the other hand, $x$ is inside $\Gamma$ but outside $\Sigma(Y_\omega, R_1)$,

$$\begin{aligned} \sigma - d_\Gamma(x) &= |x-z| + |d_\Gamma(x)| = \\ &= |x-z| + |x-\tilde{y}| \geq \\ &\geq |\tilde{y}-z|, \end{aligned}$$

which gives (A.71) via (A.70) in this case. The case in which $x$ is outside $\Gamma$ and inside $\Sigma(Y_\omega, R_1)$ does not hold: indeed, if $x$ is inside $\Sigma(Y_\omega, R_1)$, since, due to (A.64), $z$ is inside $\Gamma$, then $x$ lies, in this case, on the segment between $y_\omega$ and a point inside $\Gamma$ (that is, $z$ itself): thus, since $\Gamma$ is star-shaped with respect to $y_\omega$, $x$ is inside $\Gamma$. Let us finally consider the case in which $x$ is inside $\Sigma(Y_\omega, R_1)$ and inside $\Gamma$. In this case, note that $\tilde{y}$ cannot lie on $\Sigma(Y_\omega, R_\omega)$ (that is, $\tilde{y} \notin \Gamma_3$), otherwise $\tilde{y}$ would be on the radius from $y_\omega$ to $z$ and so

$$\tilde{y} = y_\omega + \frac{r_\omega}{r_1}(z - y_\omega)$$

which gives that

$$|\tilde{y}'| = \frac{r_\omega}{r_1}|z'| \geq \frac{r_\omega}{r_1}\eta q > \omega q,$$

if $R$ is large, thanks to (A.56), in contrast with the fact that $\tilde{y} \in \Gamma_3$.

Also, if $x$ is inside $\Sigma(Y_\omega, R_1)$ and inside $\Gamma$ and $\tilde{y}$ lies on $\Gamma_1$, then note that it must be outside $\Sigma(Y_\omega, R_1)$, thanks to (A.63) and (A.56). Thus, we take $\tilde{z} \in \Sigma(Y_\omega, R_1)$ on the segment joining $x$ and $\tilde{y}$ and we define

$$\tilde{w} := y_\omega + \frac{r_\omega}{r_1}(\tilde{z} - y_\omega).$$

Note that $\tilde{w} \in \Sigma(Y_\omega, R_\omega)$ and therefore, by (A.63),

$$|\tilde{w} - \tilde{y}| \geq \text{const}\,(1-\omega)\frac{q^2}{r}.$$

Furthermore, by (A.56),
$$\begin{aligned} |\tilde{w} - \tilde{z}| &= \left| \frac{r_\omega}{r_1}(\tilde{z} - y_\omega) - (\tilde{z} - y_\omega) \right| = \\ &= \left| \frac{r_\omega}{r_1} - 1 \right| r_1 = \\ &= |r_\omega - r_1| \leq \frac{5\,\overline{C}_0}{R}. \end{aligned}$$

Therefore,
$$\begin{aligned} \sigma - d_\Gamma(x) &= -|x - z| + |d_\Gamma(x)| \geq \\ &\geq -|x - \tilde{z}| - |x - \tilde{y}| = \\ &= |\tilde{y} - \tilde{z}| \geq \\ &\geq |\tilde{y} - \tilde{w}| - \frac{5\,\overline{C}_0}{R} \geq \\ &\geq \operatorname{const} (1-\omega) \frac{q^2}{R} - \frac{5\,\overline{C}_0}{R}, \end{aligned}$$

which proves (A.71) also in this case, by taking $q$ conveniently large (possibly in dependence of $1/(1-\omega)$ and $\overline{C}_0$).

Then, to end the proof of (A.71), we only need to consider the case in which $x$ is inside $\Sigma(Y_\omega, R_1)$ and inside $\Gamma$ and $\tilde{y}$ lies on $\Gamma_2$. For this, let us define
$$\tilde{x} = x - \frac{6\overline{C}_0}{R} \frac{x - y_\omega}{|x - y_\omega|}$$

and observe that
$$r_1 \sin\left( \angle(z - y_\omega, e_N) \right) = |z'| \geq \eta q,$$

thus, recalling (A.56),
$$\begin{aligned} \angle(\tilde{x} - y_\omega, e_N) &= \angle(x - y_\omega, e_N) = \\ &= \angle(z - y_\omega, e_N) \geq \\ &\geq \frac{\eta q}{r_1} - \operatorname{const} \frac{q^2}{r^2} \geq \\ &\geq \frac{\eta q}{r_\omega} - \operatorname{const} \frac{q^2}{r^2}. \end{aligned}$$

Also, by the minimality property of $\tilde{y}$, $x - \tilde{y}$ is orthogonal to $\Gamma$ at $\tilde{y}$ and so, by Lemma B.10, we have that $y_\omega, x, \tilde{x}, z$ and $\tilde{y}$ lie on the same plane. Let us denote this plane by $\Pi_0$. Let also $\xi_0$ be the point on the common boundary of $\Gamma_2$ and $\Gamma_3$ on the side where $\tilde{y}$ lies and let $\ell$ be the straight line lying in $\Pi_0$ and tangent to $\Gamma$ at $\xi_0$. By construction, $|\xi_0'| = \omega q$, thus
$$r_\omega \sin\left( \angle(\xi_0 - y_\omega, e_N) \right) = \omega q$$

and thence
$$\angle(\tilde{x} - y_\omega, \xi_0 - y_\omega) \geq \frac{(\eta - \omega)q}{r_\omega} - \operatorname{const} \frac{q^2}{r^2} = \frac{(1-\epsilon)(1-\omega)q}{r_\omega} - \operatorname{const} \frac{q^2}{r^2},$$

from which
$$\angle(\tilde{x} - y_\omega, \xi_0 - y_\omega) \geq \operatorname{const} \frac{(1-\omega)q}{r}.$$

## A.2. PROOF OF LEMMA 4.2

We now estimate

$$\min_{\xi \in \ell} |\xi - \tilde{x}|.$$

For this, note that the point $\xi^* \in \ell$ attaining such minimum must be so that $\tilde{x} - \xi^*$ is orthogonal to $\ell$ at $\xi^*$ and so the points $\xi_0$, $\xi^*$, $\tilde{x}$ and $y_\omega$ form a right trapezoid lying on the plane $\Pi_0$. Elementary trigonometry on this right trapezoid gives that

$$\begin{aligned}
\min_{\xi \in \ell} |\xi - \tilde{x}| &= |\xi^* - \tilde{x}| = \\
&= |\xi_0 - y_\omega| - |\tilde{x} - y_\omega| \cos\left(\angle(x - y_\omega, e_N)\right) = \\
&= r_\omega - |\tilde{x} - y_\omega| \cos\left(\angle(x - y_\omega, e_N)\right) \geq \\
&\geq r_\omega - |\tilde{x} - y_\omega| \left(1 - \frac{\text{const}\,(1-\omega)^2 q^2}{r^2}\right) \geq \\
&\geq r_\omega - |x - y_\omega| \left(1 - \frac{\text{const}\,(1-\omega)^2 q^2}{r^2}\right) - \frac{7\overline{C}_0}{R}.
\end{aligned}$$

With this estimate, we now complete the proof of (A.71) in the case in which $x$ is inside $\Sigma(Y_\omega, R_1)$ and inside $\Gamma$ and $\tilde{y}$ lies on $\Gamma_2$ by arguing as follows. Since $\tilde{x} \in \Sigma(Y_\omega, R_\omega)$ thanks to (A.56), we can take $\xi \in \ell$ on the segment joining $\tilde{y}$ to $\tilde{x}$. Then,

$$\begin{aligned}
\sigma - d_\Gamma(x) &= -|x - z| + |x - \tilde{y}| = \\
&= |x - y_\omega| - r_1 + |x - \tilde{y}| \geq \\
&\geq |x - y_\omega| - r_1 + |\tilde{x} - \tilde{y}| - \frac{6\overline{C}_0}{R} \geq \\
&\geq |x - y_\omega| - r_1 + |\tilde{x} - \xi| - \frac{6\overline{C}_0}{R} \geq \\
&\geq |x - y_\omega| - r_\omega + |\tilde{x} - \xi| - \frac{11\overline{C}_0}{R} \geq \\
&\geq |x - y_\omega| \frac{\text{const}\,(1-\omega)^2 q^2}{r^2} - \frac{18\overline{C}_0}{R}.
\end{aligned}$$

Since $x \in [-l, l]^N$, $|x - y_\omega| \geq \text{const}\, r$, therefore the above estimate yields the proof of (A.71) in the case in which $x$ is inside $\Sigma(Y_\omega, R_1)$ and inside $\Gamma$ and $\tilde{y}$ lies on $\Gamma_2$. This ends the proof of (A.71).

Then, (A.71), the fact that $R_1 \geq R/4$, (2.39) and (A.69) imply that

$$
\begin{aligned}
g_{\mathbb{S}(Y_\omega, R_1)}(x) &\geq \\
&\geq g_{y_{N+1}, R}\left(d_\Gamma(x) + \frac{3\overline{C}_0}{R} - \frac{\overline{C}_0}{2}\left(\frac{1}{R} + \frac{1}{R_1}\right)\right) = \\
&= g_{y_{N+1}, R}\Big(d_\Gamma(x) + H_{y_{N+1}, R}(0) + \frac{\overline{C}_0}{2R}y_{N+1}^2 + \\
&\quad + \frac{3\overline{C}_0}{R} - \frac{\overline{C}_0}{2}\left(\frac{1}{R} + \frac{1}{R_1}\right)\Big) \geq \\
&\geq g_{y_{N+1}, R}\left(d_\Gamma(x) + H_{y_{N+1}, R}(0) + \frac{\overline{C}_0}{2R}y_{N+1}^2 + \frac{3\overline{C}_0}{R} - \frac{5\overline{C}_0}{2R}\right) > \\
&> g_{y_{N+1}, R}\left(d_\Gamma(x) + H_{y_{N+1}, R}(0)\right) = \\
&= g_\Psi(x).
\end{aligned}
$$

This ends the proof of (A.68).

We now observe that

(A.72)
$$\mathbb{S}(Y_\omega, R_1) \cap \{|x'| \geq \check{C}q\}$$
is strictly above $\Psi$ (in the $e_{N+1}$-direction).

To prove this, take $x$ with $|x'| \geq \check{C}q$ and let $z \in \Sigma(Y_\omega, R_1)$ be realizing $d_{\Sigma(Y_\omega, R_1)}(x)$. Then, by a triangle similarity argument, one sees that

$$
\begin{aligned}
|z'| &= \frac{r_1 |x'|}{|x - y_\omega|} \geq \\
&\geq \text{const}\, \frac{r_1 \check{C} q}{l + R} \geq \\
&\geq \text{const}\, \check{C}q \geq \eta q,
\end{aligned}
$$

if $\check{C}$ is chosen appropriately large. Then, (A.72) follows from (A.68).

We now consider the domain

$$\{|x'| \leq \check{C}q\} \times \{|x_N| \leq l/2\}$$

and we slide $\Psi$ from $-\infty$ in the $e_N$ direction, until we touch $u$ for the first time in such domain (this must happen since $|u(x_0)| < 1/2$): say, for fixing the notations, that for some $\beta \in \mathbb{R}$, $\Psi - \beta e_N$ touches for the first time the graph of $u$ by above at a point $Z$. Notice that, by the hypotheses of Lemma A.1 and (A.38), we have that

(A.73) $$u(x_0) = g_{\mathbb{S}(Y, R)}(x_0) = g_\Psi(x_0),$$

therefore we have that $\beta \geq 0$. More precisely, it holds that

(A.74) $$\beta > 0.$$

Indeed, if, by contradiction, $\beta = 0$, (A.73) says that $\Psi$ touches the graph of $u$ from above at $x_0$ (which is, by construction, an interior point): thence, by (A.53), $d_\Gamma(x_0)$ must be realized on $\Gamma_1 \cup \Gamma_3$. But, by construction, $d_\Gamma(x_0)$ is realized at $T_{Y,R}(x_0) \in \Gamma_2$: this contradiction proves (A.74).

Note now that, if $\mathfrak{x}$ is in the domain of $\Psi$, then, by (2.16), we have that

$$d_\Gamma(\mathfrak{x}) \leq H_{y_{N+1}, R}(1) - H_{y_{N+1}, R}(0) \leq \text{const}\,(1 + \log R) \leq \text{const}\,(1 + \log l).$$

So, if $\mathfrak{r}_N \geq 1$, the fact that $\Gamma$ is below $\{x_N = 1\}$ implies that
$$d_\Gamma(\mathfrak{r}) \geq \mathfrak{r}_N - 1$$
and so
$$\mathfrak{r}_N \leq \operatorname{const}(1 + \log l).$$
This says that the domain of $\Psi$ is below the hyperplane
$$\left\{x_N = \operatorname{const}(1 + \log l)\right\}$$
with respect to the $e_N$-direction. Since $\beta \geq 0$, also the domain of $\Psi - \beta e_N$ is below the hyperplane
(A.75) $$\left\{x_N = \operatorname{const}(1 + \log l)\right\}.$$

What is more, since $|z_N| \leq l$, by (A.21), we gather that
(A.76) $$z_N \geq -l \geq -cR \geq y_N,$$
provided that $c$ in (A.21) is small enough. Thence, from (A.74) and (A.76),
$$\begin{aligned} |z + \beta e_N - y|^2 &= |z-y|^2 + \beta^2 + 2\beta(z_N - y_N) > \\ &> |z-y|^2, \end{aligned}$$
that is
(A.77) $$|z-y| < |z + \beta e_N - y|.$$
We now prove that
(A.78) $$|z'| < \check{C} q.$$
Indeed, if (A.78) were false, then (A.54) and (A.77) would yield that
$$\begin{aligned} u(z) &= g_\Psi(z + \beta e_N) = \\ &= g_{\mathbb{S}(Y,R)}(z + \beta e_N) = \\ &= g_{y_{N+1},R}\Big(|z + \beta e_N - y| - R + H_{y_{N+1},R}(0)\Big) > \\ &> g_{y_{N+1},R}\Big(|z - y| - R + H_{y_{N+1},R}(0)\Big) = \\ &= g_{\mathbb{S}(Y,R)}(z) \geq \\ &\geq u(z). \end{aligned}$$
This contradiction proves (A.78).

By means of (A.78), (A.75) and (A.28), we have that $z$ lies in the interior of $\{|x'| \leq \check{C} q\} \times \{|x_N| \leq l/2\}$.

We now claim that
(A.79) $$|z_{N+1}| < 1/2 \text{ and } d_\Gamma(z + \beta e_N) \text{ is realized on } \Gamma_3.$$
Indeed, since $\Psi$ touches by above $u + \beta e_N$ at $\bar{z} := z + \beta e_N$, (A.53) implies that
$$1/2 > |\bar{z}_{N+1}| = |z_{N+1}|$$

and that $d_\Gamma(\bar{z})$ is realized on $\Gamma_1 \cup \Gamma_3$. To prove (A.79), we thus have to show that $d_\Gamma(\bar{z})$ is not realized on $\Gamma_1$. If, by contradiction, $d_\Gamma(\bar{z})$ were realized on $\Gamma_1$, from (A.38) and (A.77), we would have that

$$\begin{aligned} u(z) &= g_\Psi(z + \beta e_N) = \\ &= g_{\mathbb{S}(Y,R)}(z + \beta e_N) = \\ &= g_{y_{N+1},R}\Big(|z + \beta e_N - y| - R + H_{y_{N+1},R}(0)\Big) > \\ &> g_{y_{N+1},R}\Big(|z - y| - R + H_{y_{N+1},R}(0)\Big) = \\ &= g_{\mathbb{S}(Y,R)}(z) \geq \\ &\geq u(z) \,. \end{aligned}$$

This contradiction completes the proof of (A.79).

We now claim that

(A.80) $$|Y - Y_\omega| - \beta \geq 0 \,.$$

In order to get this, first observe that if $\beta \leq 5l$, then (A.80) follows by noticing that $|Y - Y_\omega| \geq \text{const}\, r$ and by taking $l/R$ suitably small. Therefore, we restrict ourselves to the proof of (A.80) under the additional assumption that

(A.81) $$\beta \geq 5l \,.$$

For this scope, note that (A.81) implies that

$$(z + \beta e_N)_N \geq \beta - l \geq 4l \,,$$

hence, recalling (A.79),

(A.82) $$d_{\Gamma_3}(z + \beta e_N) = d_\Gamma(z + \beta e_N) > 0 \,.$$

Let us observe now that, by (A.79),

$$\begin{aligned} g_{y_{N+1},R}\Big(H_0(y_{N+1}) + |z - y| - R\Big) &= \\ = g_{\mathbb{S}(Y,R)}(z) &\geq \\ \geq u(z) &= \\ = g_\Psi(z + \beta e_N) &= \\ = g_{y_{N+1},R}\Big(H_0(y_{N+1}) + d_\Gamma(z + \beta e_N)\Big) &= \\ = g_{y_{N+1},R}\Big(H_0(y_{N+1}) + d_{\Gamma_3}(z + \beta e_N)\Big) \,, \end{aligned}$$

thence

(A.83) $$|z - y| - R \geq d_{\Gamma_3}(z + \beta e_N) \,.$$

If we now take any $x \in \Gamma_3$, we have that $x_N \leq l$ and thus

$$(z + \beta e_N - x)_N \geq \beta - 2l \,.$$

This and (A.82) imply that

$$d_{\Gamma_3}(z + \beta e_N) \geq \beta - 2l \,.$$

For this reason, recalling (A.83), we deduce that

$$\begin{aligned}\beta &\leq 2l + d_{\Gamma_3}(z + \beta e_N) \leq \\ &\leq 2l + |z - y| - R = \\ &= 2l + \operatorname{dist}(z, \partial B_r(y)) + r - R \leq \\ &\leq 2l + |z - T_{y,R}x_0| + r - R \leq \\ &\leq 5l \,.\end{aligned}$$

This ends the proof of (A.80), by taking $l/R$ suitably small.

Let us now consider, for $t \geq 0$, the surface $\mathbb{S}(Y + te_N, R_1)$. For $t = 0$, this surface is above $\mathbb{S}(Y, R)$ since, thanks to (A.55) and (2.14),

$$\begin{aligned}g_{\mathbb{S}(Y,R)}(x) &= g_{y_{N+1},R}\Big(H_0(y_{N+1}) + |x - y| - R\Big) \leq \\ &\leq g_{y_{N+1},R}\Big(H_0(y_{N+1}) + |x - y| - R_1\Big) \leq \\ &\leq g_{y_{N+1},R_1}\Big(H_0(y_{N+1}) + |x - y| - R_1\Big) = \\ &= g_{\mathbb{S}(Y,R_1)}(x)\,.\end{aligned}$$

Since $\mathbb{S}(Y, R)$ is above the graph of $u$ by our hypotheses in $\{|x'| \leq \check{C}q\} \times \{|x_N| \leq l/2\}$, we thus deduce that $\mathbb{S}(Y + te_N, R_1)$ is, for $t = 0$, above the graph of $u$ in $\{|x'| \leq \check{C}q\} \times \{|x_N| \leq l/2\}$. Hence, we may increase $t$ till we touch the graph of $u$ in $\{|x'| \leq \check{C}q\} \times \{|x_N| \leq l/2\}$. In order to fix the notation, say this happen for $t = t_1 \geq 0$ and let $X_1$ be the above mentioned touching point. Set also

$$Y_1 := Y + t_1 e_N\,.$$

Let also

(A.84) $$\tilde{r}_1 := R_1 + H_{y_{N+1},R_1}(-1/2) - H_0(y_{N+1})\,,$$

so that

$$\left\{g_{\mathbb{S}(Y_1,R_1)} = -\frac{1}{2}\right\} = \partial B_{\tilde{r}_1}(y_1)\,.$$

Then, since $u(x_0) > -1/2$ by assumption, the first touching property of $X_1$ implies that $x_0$ is above $\partial B_{\tilde{r}_1}(y_1)$. In particular, recalling (A.26), since

$$y_1 + (x_0', 0) + \sqrt{\tilde{r}_1^2 - |x_0'|^2}\, e_N \in \partial B_{\tilde{r}_1}(y_1)\,,$$

we have, using (A.24), that

(A.85) $$\begin{aligned}y_{1,N} + \sqrt{\tilde{r}_1^2 - 4q^2} &\leq y_{1,N} + \sqrt{\tilde{r}_1^2 - |x_0'|^2} \leq \\ &\leq x_{0,N} \leq \\ &\leq \operatorname{const}.\end{aligned}$$

By construction, the domain of $\mathbb{S}(Y_1, R_1)$ lies below the hyperplane

$$\left\{x_N = y_{1,N} + R_1 - H_0(y_{N+1}) + H_{y_{N+1},R_1}(1)\right\},$$

thus, (A.84), (A.85) and (2.16) imply that the domain of $\mathbb{S}(Y_1, R_1)$ lies below the hyperplane $\left\{x_N = \operatorname{const}(1 + \log R_1)\right\}$ and therefore below the hyperplane

(A.86) $$\left\{x_N = \operatorname{const}(1 + \log l)\right\}.$$

Notice also that, when $Y + te_N = Y_\omega - \beta e_N$, (A.66) and (A.79) imply that

$$\begin{aligned} g_{\mathbb{S}(Y+te_N,R_1)}(z) &= g_{\mathbb{S}(Y_\omega-\beta e_N,R_1)}(z) = \\ &= g_{\mathbb{S}(Y_\omega,R_1)}(z + \beta e_N) \leq \\ &\leq g_\Psi(z + \beta e_N) = \\ &= u(z). \end{aligned}$$

This and (A.80) say that

(A.87) $$0 \leq t_1 \leq |Y - Y_\omega| - \beta.$$

We also have that

(A.88) $$X_1 \in \{|x'| < \check{C}q\}.$$

Indeed, if (A.88) were false, then $|x'_\sharp| \geq \check{C}q$, where

(A.89) $$x_\sharp := x_1 + |Y_\omega - Y_1|e_N.$$

Thus, in the light of (A.72), we would get that

$$g_{\mathbb{S}(Y_\omega,R_1)}(x_\sharp) > g_\Psi(x_\sharp).$$

Therefore,

(A.90) $$\begin{aligned} u(x_1) &= g_{\mathbb{S}(Y_1,R_1)}(x_1) = \\ &= g_{\mathbb{S}(Y_\omega-|Y_\omega-Y_1|e_N,R_1)}(x_1) = \\ &= g_{\mathbb{S}(Y_\omega-\hat\beta e_N,R_1)}(x_1) = \\ &= g_{\mathbb{S}(Y_\omega,R_1)}(x_1 + \hat\beta e_N) = \\ &= g_{\mathbb{S}(Y_\omega,R_1)}(x_\sharp) > \\ &> g_\Psi(x_\sharp) = \\ &= g_\Psi(x_1 + \hat\beta e_N) \geq \\ &\geq u(x_1). \end{aligned}$$

This contradiction proves (A.88).

Notice also that $x_{1,N} > -l/4$, otherwise, from (A.88), (A.27), (A.55) and Lemma 2.4, we would get

$$s_{R_1} > s_R = u(x_1) = g_{\mathbb{S}(Y_1,R_1)}(x_1) \geq s_{R_1},$$

which is, of course, a contradiction.

Therefore, recalling also (A.88) and (A.86), we have that $x_1$ is in the interior of the domain $\{|x'| \leq \check{C}q\} \times \{|x_N| \leq l/2\}$, thence, by means of Proposition 2.13, $|u(x_1)| < 1/2$.

We now claim that

(A.91) $$T_{Y_1,R_1}(x_1) \in \{|x'| < \eta q\} \cap \{x_N < \operatorname{const} q^2/R\},$$

where $T_{\cdot,\cdot}$ was introduced on page 88 and $\eta$ in (A.33). To prove (A.91), first observe that, by (A.87), $y_1$ is below $y_\omega - \beta e_N$ in the $e_N$-direction, and so $\Sigma(Y_1, R_1)$ is below $\Sigma(Y_\omega - \beta e_N, R_1)$ (and, a fortiori, below $\Sigma(Y_\omega, R_1)$) in the $e_N$-direction. Thence, by (A.65),

(A.92) $$T_{Y_1,R_1}(x_1) \in \Sigma(Y_1, R_1) \subseteq \{x_N < \operatorname{const} q^2/R\}.$$

This gives a first step towards the proof of (A.91); we now show that

(A.93) $$T_{Y_1,R_1}(x_1) \in \{|x'| < \eta q\}.$$

Assume, by contradiction, that (A.93) is false. Then, by translating in the $e_N$-direction, we have that

(A.94) $$T_{Y_\omega,R_1}(x_\sharp) \in \{|x'| \geq \eta q\},$$

where $x_\sharp$ is the one defined in (A.89). Let now

$$\hat{\beta} := |Y - Y_\omega| - t_1 = \\ = |Y_\omega - Y_1|.$$

Then, $x_\sharp = x_1 + \hat{\beta} e_N$ and, in the light of (A.87), $\hat{\beta} \geq \beta$. The latter inequality and the first touching property of $\beta$ imply that $\Psi - \hat{\beta} e_N$ is above the graph of $u$, that is

(A.95) $$g_\Psi(x + \hat{\beta} e_N) \geq u(x),$$

for any $x$ in the domains of definition of the above functions. With this information, we now derive the desired contradiction. Indeed, from (A.94) and (A.68), we get that

$$g_{\mathbb{S}(Y_\omega,R_1)}(x_\sharp) > g_\Psi(x_\sharp).$$

Thence, using also the touching property of $X_1$ and (A.95), repeating the argument in (A.90) verbatim, one obtains the contradiction which ends the proof of (A.93). Thus, (A.93) and (A.92) end the proof of (A.91).

The fact that $R_1 \geq R_0/4$, (A.88) and (A.91) end the proof of (A.30) (in case $k = 0$, the other steps being analogous) and, therefore, the proof of (A.29).

With (A.29) in hand, we now complete the proof of Lemma A.1 (and this will still take some effort, the proof ending on page 123). For this, let us make some estimates on the point $x_*$ found in (A.29).

Since $x_*$ is a touching point between the barrier and $u$, we have that $|u(x_*)| \leq 1/2$ thanks to Corollary 2.14 and, therefore,

$$|\nabla u(x_*)| = |\nabla g_{\mathbb{S}(Y_*,R_*)}(x_*)| \neq 0.$$

We now show that

(A.96) $$\left| T_{Y_*,R_*} x_* - x_* + \frac{\nabla u(x_*)}{|\nabla u(x_*)|} H_0(u(x_*)) \right| \leq \frac{\text{const}}{R}.$$

Note that (A.96) is obviously fulfilled if $T_{Y_*,R_*} x_* = x_*$, since, in this case

$$H_0(u(x_*)) = H_0(g_{\widetilde{\mathbb{S}}(Y_*,R_*)}(x_*)) = H_0(g_{\widetilde{\mathbb{S}}(Y_*,R_*)}(T_{Y_*,R_*} x_*)) = H_0(0) = 0.$$

Hence, we focus on the proof of (A.96) under the assumption $T_{Y_*,R_*} x_* \neq x_*$. For this scope, first notice that $x_* - y_*$ is, by construction, parallel to $x_* - T_{Y_*,R_*} x_*$, that is

(A.97) $$\frac{x_* - y_*}{|x_* - y_*|} = \pm \frac{x_* - T_{Y_*,R_*} x_*}{|x_* - T_{Y_*,R_*} x_*|},$$

where the sign $+/-$ takes into account the case in which $x_*$ is outside/inside $\Sigma(Y_*, R_*)$. Also, from (A.19),

(A.98) $$\left| x_* - T_{Y_*,R_*} x_* \right| \leq \text{const}.$$

In the light of (2.57) and (A.97),

$$
\text{(A.99)} \qquad \frac{\nabla g_{\widetilde{\mathbb{S}}(Y_*,R_*)}(x_*)}{|\nabla g_{\widetilde{\mathbb{S}}(Y_*,R_*)}(x_*)|} = \frac{x_* - y_*}{|x_* - y_*|} = \pm \frac{x_* - T_{Y_*,R_*}x_*}{|x_* - T_{Y_*,R_*}x_*|}.
$$

Let us now define $W(x) := H_0(g_{\widetilde{\mathbb{S}}(Y_*,R_*)}(x))$. By the touching property of $x_*$, the fact that $T_{Y_*,R_*}x_* \in \Sigma(Y_*,R_*)$ and (A.99), we have that

$$
\left| T_{Y_*,R_*}x_* - x_* + \frac{\nabla u(x_*)}{|\nabla u(x_*)|} H_0(u(x_*)) \right| =
$$

$$
= \left| T_{Y_*,R_*}x_* - x_* + \frac{\nabla g_{\widetilde{\mathbb{S}}(Y_*,R_*)}(x_*)}{|\nabla g_{\widetilde{\mathbb{S}}(Y_*,R_*)}(x_*)|} W(x_*) \right| =
$$

$$
= \left| \frac{\nabla g_{\widetilde{\mathbb{S}}(Y_*,R_*)}(x_*)}{|\nabla g_{\widetilde{\mathbb{S}}(Y_*,R_*)}(x_*)|} \left[ \mp |T_{Y_*,R_*}x_* - x_*| + W(x_*) \right] \right| =
$$

$$
= \left| \mp |T_{Y_*,R_*}x_* - x_*| + W(x_*) \right| =
$$

$$
= \left| \mp |T_{Y_*,R_*}x_* - x_*| + W(x_*) - W(T_{Y_*,R_*}x_*) \right|.
$$

Therefore, a second order Taylor expansion of $W$, (2.54), (A.98), (A.99) and (2.53) give

$$
\left| T_{Y_*,R_*}x_* - x_* + \frac{\nabla u(x_*)}{|\nabla u(x_*)|} H_0(u(x_*)) \right| \le
$$

$$
\le \left| \mp |T_{Y_*,R_*}x_* - x_*| - \right.
$$

$$
\left. - \nabla W(x_*)(T_{Y_*,R_*}x_* - x_*) \right| +
$$

$$
+ \text{const}\, |D^2 W|\, |T_{Y_*,R_*}x_* - x_*|^2 \le
$$

$$
\le \left| \mp |T_{Y_*,R_*}x_* - x_*| - \right.
$$

$$
\left. - H_0'(g_{\widetilde{\mathbb{S}}(Y_*,R_*)}(x_*)) \nabla g_{\widetilde{\mathbb{S}}(Y_*,R_*)}(x_*) \cdot (T_{Y_*,R_*}x_* - x_*) \right| +
$$

$$
+ \frac{\text{const}}{R} \le
$$

$$
\le \left| \frac{T_{Y_*,R_*}x_* - x_*}{|T_{Y_*,R_*}x_* - x_*|} \left[ T_{Y_*,R_*}x_* - x_* - \right.\right.
$$

$$
\left.\left. - H_0'(g_{\widetilde{\mathbb{S}}(Y_*,R_*)}(x_*)) |\nabla g_{\widetilde{\mathbb{S}}(Y_*,R_*)}(x_*)| \cdot (T_{Y_*,R_*}x_* - x_*) \right] \right| +
$$

$$
+ \frac{\text{const}}{R} =
$$

$$
= \left| T_{Y_*,R_*}x_* - x_* - \right.
$$

$$
\left. - H_0'(g_{\widetilde{\mathbb{S}}(Y_*,R_*)}(x_*)) |\nabla g_{\widetilde{\mathbb{S}}(Y_*,R_*)}(x_*)| \cdot (T_{Y_*,R_*}x_* - x_*) \right| +
$$

$$
+ \frac{\text{const}}{R} \le
$$

$$
\le \left| T_{Y_*,R_*}x_* - x_* - (T_{Y_*,R_*}x_* - x_*) \right| + \frac{\text{const}}{R},
$$

that is (A.96).

We now show that
$$\angle\left(\frac{\nabla u(x_*)}{|\nabla u(x_*)|}, e_N\right) \leq \frac{\text{const } q}{R}. \tag{A.100}$$

For proving this, let us note that
$$|T_{Y_*,R_*}x_* - y_*| = r(Y_*, R_*) =$$
$$= R_* - H_0(y_{N+1}) - \frac{\overline{C_0}}{2R}y_{N+1}^2 \geq \tag{A.101}$$
$$\geq \text{const } R_*,$$

hence,
$$\sin[\angle(T_{Y_*,R_*}x_* - y_*, e_N)] =$$
$$= \frac{|(T_{Y_*,R_*}x_* - y_*)'|}{|T_{Y_*,R_*}x_* - y_*|} =$$
$$= \frac{|(T_{Y_*,R_*}x_*)'|}{|T_{Y_*,R_*}x_* - y_*|} \leq \tag{A.102}$$
$$\leq \frac{\text{const } q}{R}.$$

Also, in analogy with (A.99), we have that
$$\frac{\nabla u(x_*)}{|\nabla u(x_*)|} = \frac{g_{\widetilde{\mathbb{S}}(Y_*,R_*)}(x_*)}{|g_{\widetilde{\mathbb{S}}(Y_*,R_*)}(x_*)|} = \frac{T_{Y_*,R_*}x_* - y_*}{|T_{Y_*,R_*}x_* - y_*|},$$

which, together with (A.102), proves (A.100).

We now observe that, thanks to (A.96) and (A.100), we have
$$x_* \cdot e_N \leq$$
$$\leq (T_{Y_*,R_*}x_*) \cdot e_N + \frac{\nabla u(x_*)}{|\nabla u(x_*)|} \cdot e_N\, H_0(u(x_*)) + \frac{\text{const}}{R} \leq \tag{A.103}$$
$$\leq (T_{Y_*,R_*}x_*) \cdot e_N + H_0(u(x_*)) + \frac{\text{const } q^2}{R^2} + \frac{\text{const}}{R}.$$

Let us now consider the set
$$\Xi_* := \left\{\bar Y = (\bar y', 0, \bar y_{N+1}) \in \mathbb{R}^{N+1} \text{ s.t.} \right. \tag{A.104}$$
$$\left. |(\bar y - T_{Y_*,R_*}x_*)'| \leq cq,\ |\bar y_{N+1}| \leq 1/4 \right\}$$

and let
$$\check R := cR \tag{A.105}$$

with $c > 0$ suitably small. Recalling the definition of $r(\cdot, \cdot)$ given in (A.18), we also denote $\check r := r(\check Y, \check R)$.

For any $\bar Y \in \Xi_*$, let us slide $\mathbb{S}(\bar Y, \check R)$ from $-\infty$ in the $e_N$ direction, until it touches the graph of $u$ by above for the first time, and let $\check Y$ denote the center of the corresponding barrier: more explicitly, $\check Y = \bar Y - te_N$, for some $t \in \mathbb{R}$ and $\mathbb{S}(\check Y, \check R)$ touches the graph of $u$ from above for the first time coming from $-\infty$ in the $e_N$ direction. Say, also, to fix notations, that such touching occurs at some point $\check{X}$. We will denote by $\tilde{\Xi}$ the set collecting all the points $\check{X}$ which lie in the interior of the domain of $u$, when $\bar{Y}$ varies in $\Xi_*$.

We know from (A.29) that

(A.106) $$\left(T_{Y_*,R_*}x_*\right)_N \leq C_\sharp \frac{q^2}{R}$$

for some suitably large constant $C_\sharp$. We now claim that, if $\bar{Y} \in \Xi_*$, then

(A.107) $$\left(\left(T_{Y_*,R_*}x_*\right)', 2C_\sharp \frac{q^2}{R}\right) \text{ is outside } \Sigma(\check{Y}, \check{R}).$$

The proof of (A.107) (which is pretty long and will be completed only on page 117) is by contradiction. If (A.107) were false, Lemma B.7 would imply that

(A.108) $$\Sigma(\check{Y}, \check{R}) \cap \{|(x - T_{Y_*,R_*}x_*)'| \leq cq\} \cap \{x_N \geq \check{y}_N\}$$
is above the hyperplane $\{x_N = 3C_\sharp q^2/(2R)\}$,

provided that $c$ is small enough[7]. We now show that

(A.109) $$d_{\Sigma(\check{Y},\check{R})}(x_*) \leq$$
$$\leq H_0(u(x_*)) + \text{const } \frac{q}{R} - \frac{C_\sharp q^2}{2R}.$$

To prove this (and some effort will be needed), we distinguish two cases: either $x_*$ is in the exterior or it is in the interior of $\Sigma(\check{Y}, \check{R})$.

Let us first assume that $x_*$ is not in the interior of $\Sigma(\check{Y}, \check{R})$. We first point out that

(A.110) $$x_* \cdot e_N \geq \frac{3C_\sharp q^2}{2R}.$$

For this, note that, by (A.98),

$$\left|(x_* - T_{Y_*,R_*}x_*)'\right| \leq cq$$

if $q$ is large enough. This, (A.108) and the assumption that $x_*$ is not in the interior of $\Sigma(\check{Y}, \check{R})$ would imply that either (A.110) holds or $x_*$ is below $\Sigma(\check{Y}, \check{R})$. Thus, to complete the proof of (A.110), we show now that the latter possibility cannot hold. Indeed, note that the first touching property of $\mathbb{S}(\check{Y}, \check{R})$ and Lemma 2.25 imply that

$$\check{y} \cdot e_N \leq \check{x} \cdot e_N$$

while (A.19) says that

$$|\check{x} - \check{y}| \geq \text{const } \check{R}.$$

Furthermore, in analogy with (A.100), we have that

(A.111) $$\angle(\check{x} - \check{y}, e_N) \leq \frac{\text{const } q}{R}$$

---

[7]The reader will indeed notice that, due to (A.104),

$$\check{y} + re_N \in \Sigma(\check{Y}, \check{R}) \cap \{|(x - T_{Y_*,R_*}x_*)'| \leq cq\} \cap \{x_N \geq y_N\} \neq \emptyset.$$

Then, the above estimates yield that

$$(x_* - \breve{y}) \cdot e_N \geq (\breve{x} - \breve{y}) \cdot e_N - \operatorname{const} l =$$
$$= |(\breve{x} - \breve{y}) \cdot e_N| - \operatorname{const} l \geq$$
(A.112)
$$\geq \frac{1}{2} |\breve{x} - \breve{y}| - \operatorname{const} l \geq$$
$$\geq \operatorname{const} \breve{R} - \operatorname{const} l,$$

thus showing that $(x_* - \breve{y}) \cdot e_N > 0$ and thus that $x_*$ is not below $\Sigma(\breve{Y}, \breve{R})$. This ends the proof of (A.110).

With this, we now go back to the proof of (A.109) when $x_*$ is not in the interior of $\Sigma(\breve{Y}, \breve{R})$. By means of (A.108), we have that the point $\left((T_{Y_*,R_*} x_*)', 3C_\sharp q^2/(2R)\right)$ lies inside $\Sigma(\breve{Y}, \breve{R})$, thus

$$d_{\Sigma(\breve{Y}, \breve{R})}(x_*) \leq \left| x_* - \left((T_{Y_*,R_*} x_*)', \frac{3C_\sharp q^2}{2R}\right)\right|,$$

and therefore, recalling (A.110),

$$d_{\Sigma(\breve{Y}, \breve{R})}(x_*) \leq |(x_* - T_{Y_*,R_*} x_*)'| + x^* \cdot e_N - \frac{3C_\sharp q^2}{2R}.$$

We recall that, by (A.100),

$$\angle\left((x_* - T_{Y_*,R_*}), e_N\right) = \angle\left(\frac{\nabla u(x_*)}{|\nabla u(x_*)|}, e_N\right) \leq \frac{\operatorname{const} q}{R},$$

and so, by means of (A.98), we have that

(A.113)
$$\left|((T_{Y_*,R_*} x_*) - x_*)'\right| \leq \operatorname{const} \frac{q}{R}.$$

Therefore, recalling (A.98) and taking into account (A.103),

$$d_{\Sigma(\breve{Y}, \breve{R})}(x_*) \leq \operatorname{const} \frac{q^2}{R} + T_{Y_*,R_*} x_* \cdot e_N + H_0(u(x_*)) - \frac{3C_\sharp q^2}{2R}.$$

Thence, if $x_*$ is not in the interior of $\Sigma(\breve{Y}, \breve{R})$, (A.109) follows by (A.106) and by taking $C_\sharp$ suitably large.

If otherwise $x_*$ is in the interior of $\Sigma(\breve{Y}, \breve{R})$, that is, if

(A.114)
$$d_{\Sigma(\breve{Y}, \breve{R})}(x_*) < 0,$$

in order to prove (A.109), we argue as follows. We first note that, in this case,

$$u(x_*) \leq g_{\widetilde{\mathbb{S}}(\breve{Y}, \breve{R})}(x_*) < 0$$

and consequently

$$g_{\widetilde{\mathbb{S}}(Y_*, R_*)}(x_*) = u(x_*) < 0,$$

so that $x_*$ is also in the interior of $\Sigma(Y_*, R_*)$. Therefore, recalling (A.100), it follows that

(A.115)
$$x_* \cdot e_N < (T_{Y_*,R_*} x_*) \cdot e_N.$$

We now deduce that, when $x_*$ is in the interior of $\Sigma(\breve{Y}, \breve{R})$, one has that

(A.116)
$$-d_{\Sigma(\breve{Y}, \breve{R})}(x_*) \geq \min\left\{\frac{cq}{4}, \frac{3C_\sharp q^2}{2R} - x_* \cdot e_N\right\}.$$

To prove (A.116), take $x \in \Sigma(\check{Y}, \check{R})$. Obviously, we may and do assume that
(A.117)
$$|x| \leq \text{const}\, l\,,$$
otherwise
$$|x_* - x| \geq \text{const}\, l$$
and so (A.116) trivially follows from (A.114). There are now two possibilities: either $x$ is above the hyperplane $\{x_N = 3C_\sharp q^2/(2R)\}$ or the converse. If $x$ is above the hyperplane $\{x_N = 3C_\sharp q^2/(2R)\}$, then
$$|x - x_*| \geq x_N - x_* \cdot e_N \geq \frac{3C_\sharp q^2}{2R} - x_* \cdot e_N\,,$$
which, together with (A.114), proves (A.116) in case $x$ is above the hyperplane $\{x_N = 3C_\sharp q^2/(2R)\}$. If, on the other hand, $x$ is not above the hyperplane $\{x_N = 3C_\sharp q^2/(2R)\}$, by (A.108), we have that either $|(x - T_{Y_*, R_*} x_*)'| \geq cq$ or $x_N \leq \check{y}_N$. However, the latter cannot hold, since, by Lemma 2.16, (A.117), (A.111) and (A.112), we have that
$$\begin{aligned} x_N - \check{y}_N &= |\check{x}_N - \check{y}_N| + x_N - \check{x}_N \geq \\ &\geq \frac{1}{2}|\check{x} - \check{y}| + x_N - \check{x}_N \geq \\ &\geq \text{const}\, R - \text{const}\, l > 0\,. \end{aligned}$$
Therefore, from these considerations, we have that if $x$ is not above the hyperplane $\{x_N = 3C_\sharp q^2/(2R)\}$, then $|(x - T_{Y_*, R_*} x_*)'| \geq cq$. For this reason, recalling (A.113),
$$\begin{aligned} |(x - x_*)'| &\geq \left|(x - (T_{Y_*, R_*} x_*))'\right| - \left|((T_{Y_*, R_*} x_*) - x_*)'\right| \geq \\ &\geq cq - \frac{\text{const}\, q}{R} \geq \frac{cq}{2}\,, \end{aligned}$$
which, together with (A.115) and (A.106), yields the proof of (A.116) also when $x$ is not above the hyperplane $\{x_N = 3C_\sharp q^2/(2R)\}$. This ends the proof of (A.116).

We now exploit (A.103), (A.106) and (A.116) to get that
$$-d_{\Sigma(\check{Y}, \check{R})}(x_*) \geq -H_0(u(x_*)) - \text{const}\,\frac{q}{R} + \frac{C_\sharp q^2}{2R}\,.$$
This ends the proof of (A.109) also in the case in which $x_*$ is in the interior of $\Sigma(\check{Y}, \check{R})$.

Having completed the proof of (A.109), we now deduce the contradiction that will finish the proof of (A.107). To this end, note that, by (A.19), letting $s_* := u(x_*)$, we have that the signed distance between the $s_*$-level set of $g_{\mathbb{S}(\check{Y}, \check{R})}$ and $\Sigma(\check{Y}, \check{R})$ is bigger than
$$H_0(u(x_*)) - \frac{\text{const}}{R}\,.$$
Thus, from (A.109), by taking $q$ and $C_\sharp$ suitably large, we have that
$$d_{\Sigma(\check{Y}, \check{R})}(\mathfrak{x}) > d_{\Sigma(\check{Y}, \check{R})}(x_*)\,,$$
for any $\mathfrak{x}$ so that $g_{\mathbb{S}(\check{Y}, \check{R})}(\mathfrak{x}) = s_*$. Therefore, $x_*$ is strictly in the interior of the $s_*$-level set of $g_{\mathbb{S}(\check{Y}, \check{R})}$, that is
$$g_{\mathbb{S}(\check{Y}, \check{R})}(x_*) < s_* = u(x_*)\,.$$

This contradicts the fact that $g_{\mathbb{S}(\check{Y},\check{R})}$ touches $u$ by above and, thus, yields the proof of (A.107).

In the light of (A.107), some elementary trigonometry implies that $y + \check{r}e_N$ is below the hyperplane $\{x_N = 4C_\sharp q^2/R\}$ and therefore

(A.118) $\quad\quad\quad\quad \Sigma(\check{Y},\check{R})$ is below $\{x_N = 4C_\sharp q^2/R\}$.

Let us now fix a small constant $c_* > 0$. If $q/R$ is assumed to be small enough (possibly in dependence of $c_*/C_\sharp$), then (A.107) and Corollary B.9 imply that

(A.119) $\quad\quad\quad\quad \Sigma(\check{Y},\check{R}) \cap \{|x'| \geq c_* q\}$ is below $\{x_N = 0\}$.

We now show that the above geometric considerations imply that

(A.120) $\quad\quad$ outside $\{|x'| \leq c_* q\} \times \{x_N > 0\}$,
$\Sigma(\check{Y},\check{R})$ is at distance greater than $q^2/(4R)$
in the interior of $\Sigma(Y,R)$.

For the proof of (A.120), first notice that

(A.121) $\quad\quad\quad\quad \check{y}_N - \check{r} \geq -\mathrm{const}\, \check{r} \geq y_N$

if $c$ in (A.105) is chosen suitably small. Let us now consider the surface

$$\Sigma_t(\check{Y},\check{R}) := \Sigma(\check{Y},\check{R}) + te_N$$

with $t \geq 0$. By (A.121), it follows that, given $\mathfrak{x} \in \Sigma(\check{Y},\check{R})$, then $\mathfrak{x}_N \geq y_N$ and so

$$|(\mathfrak{x} + te_N) - y| \geq |\mathfrak{x} - y|.$$

Hence,

(A.122) $\quad\quad\quad\quad d_{\Sigma(Y,R)}(\mathfrak{x}) \leq d_{\Sigma(Y,R)}(\mathfrak{x} + te_N)$.

Recalling (A.119), we now select the first $t \geq 0$ for which

(A.123) $\quad\quad\quad\quad \mathfrak{S} := \Sigma_t(\check{Y},\check{R}) \cap \{|x'| = c_* q\} \cap \{x_N = 0\} \neq \emptyset$.

We also denote by $\mathfrak{S}^-$ the portion of $\Sigma_t(\check{Y},\check{R})$ which is below $\{x_N = 0\}$ (with respect to the $e_N$-direction), that is the portion of $\Sigma_t(\check{Y},\check{R})$ which is below $\mathfrak{S}$. The choice in (A.123) implies that

(A.124) $\quad\quad\quad\quad \check{y}_N + t + \check{r} \geq 0$

and, by (A.104) and (A.29), that

(A.125) $\quad\quad\begin{aligned} |(\check{y} + te_N)'| &= |\check{y}'| \leq \\ &\leq |(\check{y} - T_{Y_*,R_*}x_*)'| + |(T_{Y_*,R_*}x_*)'| \leq \\ &\leq cq + \frac{q}{C_*} = \\ &= \tilde{c}_* q, \end{aligned}$

where

(A.126) $\quad\quad\quad\quad \tilde{c}_* := c + \frac{1}{C_*}$

is a positive constant, which is small if $c$ and $1/C_*$ are small.

We now show that, with this construction and taking $\tilde{c}_*$ small enough, we have that $\breve{y} + te_N$ is above the cone $\mathfrak{C}_*$ defined as

$$\mathfrak{C}_* := \left\{ x \in \mathbb{R}^N \text{ s.t. } |x'| = \frac{c_* q}{\sqrt{r^2 - q^2}} |x_N + \sqrt{r^2 - q^2}| \right\}.$$

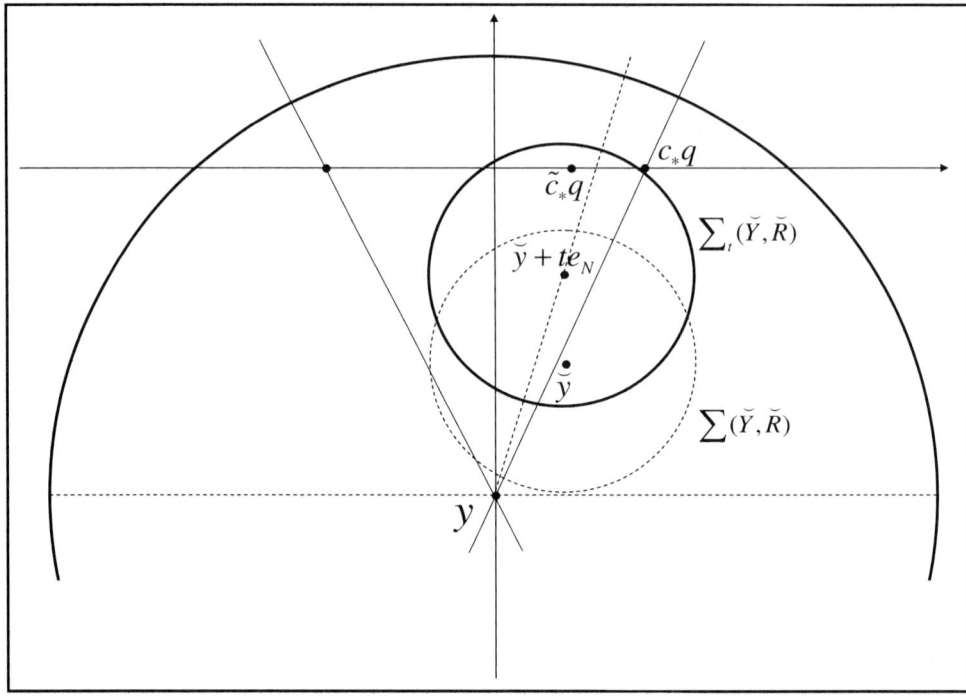

**The proof of (A.120)**

Indeed, by (A.124) and (A.125),

$$\frac{c_* q}{\sqrt{r^2 - q^2}} \left( \breve{y}_N + t + \sqrt{r^2 - q^2} \right) \geq$$
$$\geq \frac{c_* q}{\sqrt{r^2 - q^2}} \left( \sqrt{r^2 - q^2} - \breve{r} \right) =$$
$$= c_* q \left( 1 - \frac{\breve{r}}{\sqrt{r^2 - q^2}} \right) \geq$$
$$\geq \frac{c_* q}{2}$$
$$> \tilde{c}_* q \geq \left| (\breve{y} + te_N)' \right|$$

if $q/R$, $\breve{r}/r$, $c$ and $1/C_*$ (and thus $\tilde{c}_*$) are small enough (recall (A.105) and (A.126)).

This fact, (A.123) and some elementary geometric considerations yield that if $x \in \mathfrak{S}^-$, and $x_\sharp \in \mathfrak{S}$, then

(A.127) $$\left| d_{\Sigma(Y,R)}(x) \right| \geq \left| d_{\Sigma(Y,R)}(x_\sharp) \right|.$$

By construction, since $c_* < 1$, if $x_\sharp \in \mathfrak{S}$, then $x_\sharp$ lies inside $\Sigma(Y, R)$ (see the figure on page 89); thus,

(A.128) $$d_{\Sigma(Y,R)}(x_\sharp) < 0.$$

Also, from (A.123) and the fact that $\check{r} < r$, if $x \in \mathfrak{S}^-$, then $x$ is also inside $\Sigma(Y,R)$ and therefore

(A.129) $$d_{\Sigma(Y,R)}(x) < 0.$$

Thus, from (A.127), (A.128) and (A.129),

(A.130) $$d_{\Sigma(Y,R)}(x) \leq d_{\Sigma(Y,R)}(x_\sharp),$$

for $x \in \mathfrak{S}^-$ and $x_\sharp \in \mathfrak{S}$.

Also, since $y' = 0$ by its definition, we gather that, if $x_\sharp \in \mathfrak{S}$, then
$$|x_\sharp - y|^2 = |y_N|^2 + |x'_\sharp|^2 = r^2 - q^2 + c_*^2 q^2$$
and therefore

(A.131) $$d_{\Sigma(Y,R)}(x_\sharp) = |x_\sharp - y| - r = \sqrt{r^2 - q^2 + c_*^2 q^2} - r.$$

Thus, taking $x \in \mathfrak{S}^-$ and $x_\sharp \in \mathfrak{S}$, making use of (A.130) and (A.131), and recalling also (A.35), we have that

(A.132) $$\begin{aligned}-d_{\Sigma(Y,R)}(x) &\geq -d_{\Sigma(Y,R)}(x_\sharp) = \\ &= r - r\sqrt{1 - \frac{q^2}{r^2} + \frac{c_*^2 q^2}{r^2}} \geq \\ &\geq r - r\left[1 - \frac{q^2}{2r^2} + \frac{c_*^2 q^2}{2r^2} + \left(-\frac{q^2}{r^2} + \frac{c_*^2 q^2}{r^2}\right)^2\right] \geq \\ &\geq \frac{q^2}{2r}\left(1 - c_*^2 - \text{const}\,\frac{q^2}{R^2}\right) > \\ &> \frac{q^2}{4R},\end{aligned}$$

provided that $c_*$ and $q/R$ are sufficiently small.

To complete the proof of (A.120), take now $\mathfrak{x} \in \Sigma(\check{Y}, \check{R})$ outside $\{|x'| \leq c_* q\} \times \{x_N > 0\}$ and let $x := \mathfrak{x} + t e_N$. Then, $x \in \mathfrak{S}^-$ by the choice in (A.123), thus, from (A.122) and (A.132), we gather that

$$-d_{\Sigma(Y,R)}(\mathfrak{x}) \geq -d_{\Sigma(Y,R)}(x) > \frac{q^2}{4R}.$$

This ends the proof of (A.120).

Let us now prove that

(A.133) the $s$-level surface of $g_{\mathbb{S}(Y,R)}$ is at distance greater than $H_{y_{N+1},\check{R}}(s)$ from $\Sigma(Y,R)$.

In order to prove this, take any $\hat{x}$ in the $s$-level surface of $g_{\mathbb{S}(Y,R)}$, that is assume that
$$g_{y_{N+1},R}(H_0(y_{N+1}) + |\hat{x} - y| - R) = s.$$
Let also $\bar{x} \in \Sigma(Y, R)$, that is
$$g_{y_{N+1},R}(H_0(y_{N+1}) + |\bar{x} - y| - R) = 0,$$

and assume that $\bar{x}$ lies on the half-line from $y$ towards $\hat{x}$. Then,

$$\begin{aligned}
d_{\Sigma(Y,R)}(\hat{x}) &= |\hat{x}-y|-|\bar{x}-y| = \\
&= (H_0(y_{N+1})+|\hat{x}-y|-R)-(H_0(y_{N+1})+|\bar{x}-y|-R) = \\
&= H_{y_{N+1},R}(s)-H_{y_{N+1},R}(0).
\end{aligned}$$

Hence, by the fact that $\check{R} \leq R$, (2.14) and (2.39), we get that

$$\begin{aligned}
d_{\Sigma(Y,R)}(\hat{x}) &\geq H_{y_{N+1},\check{R}}(s)-H_{y_{N+1},R}(0) = \\
&= H_{y_{N+1},\check{R}}(s)+\frac{\overline{C}_0}{2R}y_{N+1}^2 \geq \\
&\geq H_{y_{N+1},\check{R}}(s),
\end{aligned}$$

and this ends the proof of (A.133).

We now show that

(A.134) if $|s| \leq 1/2$, the $s$-level surface of $g_{\mathbb{S}(\check{Y},\check{R})}$ is at distance less than $H_{y_{N+1},\check{R}}(s) + \text{const}\,\overline{C}_0/(2R)$ from $\Sigma(\check{Y},\check{R})$.

Indeed, we argue as in the proof of (A.133), by taking now $\hat{x}$ in the $s$-level surface of $g_{\mathbb{S}(\check{Y},\check{R})}$ and $\bar{x} \in \Sigma(\check{Y},\check{R})$, with $\bar{x}$ lying on the half-line from $y$ towards $\hat{x}$, and by arguing as above, we have that

$$\begin{aligned}
d_{\Sigma(\check{Y},\check{R})}(\hat{x}) &= |\hat{x}-\check{y}|-|\bar{x}-\check{y}| = \\
&= (H_0(y_{N+1})+|\hat{x}-\check{y}|-\check{R})-(H_0(y_{N+1})+|\bar{x}-\check{y}|-\check{R}) = \\
&= H_{\check{y}_{N+1},\check{R}}(s)-H_{\check{y}_{N+1},\check{R}}(0) = \\
&= H_{\check{y}_{N+1},\check{R}}(s)+\frac{\overline{C}_0}{2\check{R}}\check{y}_{N+1}^2 \leq \\
&\leq H_{\check{y}_{N+1},\check{R}}(s)+\text{const}\,\frac{\overline{C}_0}{2R}.
\end{aligned}$$

By the assumption that $|s| \leq 1/2$ and (2.39), we may now estimate the quantity $H_{\check{y}_{N+1},\check{R}}(s)$ here below with $H_{y_{N+1},\check{R}}(s)+\overline{C}_0/R$, and this ends the proof of (A.134).

We now deduce from the above estimates that

(A.135) at any $x$ for which
$|g_{\mathbb{S}(\check{Y},\check{R})}(x)| \leq 1/2$ and
$d_{\Sigma(\check{Y},\check{R})}(x) - d_{\Sigma(Y,R)}(x) \geq \text{const}\,\overline{C}_0/R$,
we have that
$g_{\mathbb{S}(\check{Y},\check{R})}(x) > g_{\mathbb{S}(Y,R)}(x)$.

To prove this, take $x$ as in (A.135) here above and let

(A.136) $$\begin{aligned}\check{s} &:= g_{\mathbb{S}(\check{Y},\check{R})}(x) \in [-1/2,\,1/2] \\ s &:= g_{\mathbb{S}(Y,R)}(x).\end{aligned}$$

Then, by (A.133) and (A.134),

(A.137) $$\begin{aligned}d_{\Sigma(\check{Y},\check{R})}(x) &< H_{y_{N+1},R}(\check{s}) + \text{const}\,\overline{C}_0/R \qquad \text{and} \\ d_{\Sigma(Y,R)}(x) &> H_{y_{N+1},R}(s) - \text{const}\,\overline{C}_0/R,\end{aligned}$$

hence, for the assumption in (A.135),

$$H_{y_{N+1},R}(s) < H_{y_{N+1},R}(\check{s}),$$

which proves (A.135) via the monotonicity of $H_{y_{N+1},R}$ and (A.136).

Next, notice that, thanks to (A.120) and (A.135), we have that

(A.138)
$$\mathbb{S}(\check{Y},\check{R}) \cap \{|x'| > c_*q\}$$
is strictly above $\mathbb{S}(Y,R)$,

provided that $q$ is large enough. Hence, since $u \leq g_{\mathbb{S}(Y,R)}$, (A.138) implies that

(A.139)
$$\mathbb{S}(\check{Y},\check{R}) \cap \{|x'| > c_*q\}$$
is strictly above the graph of $u$.

This also implies that $\check{x}$ is an interior contact point.

We now apply the previous considerations to deduce some properties of the contact points $\check{X}$ and of the contact point set $\check{\Xi}$ (recall the notation on page 114). First of all,

(A.140) $$|u(\check{x})| < 1/2,$$

thanks to Corollary 2.14. This and (A.139) imply that

(A.141) $$|\check{x}| \leq c_*q,$$

while (A.118) yields

(A.142) $$T_{\check{Y},\check{R}}\check{x} \in \Sigma(\check{Y},\check{R}) \subseteq \{x_N < 4C_\sharp q^2/R\}.$$

Also, from (A.19) and (A.140), we get that

$$\Big||\check{x} - \check{y}| - r\Big| \leq \text{const},$$

and, consequently, $|\check{x} - \check{y}| \geq r/2$, if $r$ is large. From this circumstance, recalling the touching properties of $\check{x}$, (2.57), and (A.141), we gather that

$$\sin\left(\angle\Big(\frac{\nabla u(\check{x})}{|\nabla u(\check{x})|}, e_N\Big)\right) = \sin\left(\angle\Big(\frac{\nabla g_{\mathbb{S}(\check{Y},\check{R})}(\check{x})}{|\nabla g_{\mathbb{S}(\check{Y},\check{R})}(\check{x})|}, e_N\Big)\right) =$$

$$= \sin\left(\angle\Big(\frac{\check{x} - \check{y}}{|\check{x} - \check{y}|}, e_N\Big)\right) =$$

$$= \frac{|\check{x}' - \check{y}'|}{|\check{x} - \check{y}|} \leq$$

$$\leq \text{const}\,\frac{q}{r}.$$

Thence,

(A.143) $$\angle\left(\frac{\nabla u(\check{x})}{|\nabla u(\check{x})|}, e_N\right) \leq \text{const}\,\frac{q}{R}.$$

Analogously, if $x_0$ is the point in the statement of Lemma A.1, one sees that

(A.144) $$\angle\left(\frac{\nabla u(x_0)}{|\nabla u(x_0)|}, e_N\right) \leq \text{const}\,\frac{q}{R}.$$

In particular, from (A.143) and (A.144) it follows that

(A.145) $$\angle\left(\frac{\nabla u(\check{x})}{|\nabla u(\check{x})|}, \frac{\nabla u(x_0)}{|\nabla u(x_0)|}\right) \leq \text{const}\,\frac{q}{R}.$$

In addition, from Proposition 3.14,

(A.146) $$\mathfrak{L}^N\left(\pi_N(\check{\Xi})\right) \geq \text{const}\,\mathfrak{L}^N(\Xi_*) \geq \text{const}(cq)^{N-1}.$$

What is more, arguing as in the proof of (A.96), one can see that

(A.147) $$\left|T_{\check{Y},\check{R}}\check{x} - \check{x} + \frac{\nabla u(\check{x})}{|\nabla u(\check{x})|} H_0(u(\check{x}))\right| \leq \frac{\text{const}}{R} \quad \text{and}$$
$$\left|T_{Y,R}x_0 - x_0 + \frac{\nabla u(x_0)}{|\nabla u(x_0)|} H_0(u(x_0))\right| \leq \frac{\text{const}}{R}.$$

Hence, from (A.147), (A.142) and the fact that $T_{Y,R}x_0 \cdot e_N = 0$ (recall the assumptions in Lemma A.1), we have that

$$(\check{x} - x_0)\cdot e_N \leq \Big(T_{\check{Y},\check{R}}\check{x} - T_{Y,R}x_0 +$$
$$+ \frac{\nabla u(\check{x})}{|\nabla u(\check{x})|} H_0(u(\check{x})) - \frac{\nabla u(x_0)}{|\nabla u(x_0)|} H_0(u(x_0))\Big)\cdot e_N +$$
$$+ \frac{\text{const}}{R} \leq$$
$$\leq \Big(\frac{\nabla u(\check{x})}{|\nabla u(\check{x})|} H_0(u(\check{x})) - \frac{\nabla u(x_0)}{|\nabla u(x_0)|} H_0(u(x_0))\Big)\cdot e_N +$$
$$+ \frac{4\,C_\sharp\,q^2}{R} + \frac{\text{const}}{R},$$

and so, thanks to (A.140), (A.143) and (A.144),

(A.148) $$(\check{x} - x_0)\cdot e_N \leq$$
$$\leq H_0(u(\check{x})) - H_0(u(x_0)) + \frac{\text{const}\,q^2}{R^2} + \frac{4\,C_\sharp\,q^2}{R} + \frac{\text{const}}{R} \leq$$
$$\leq H_0(u(\check{x})) - H_0(u(x_0)) + \frac{5\,C_\sharp\,q^2}{R},$$

by assuming $C_\sharp$, $q$ and $R/q$ suitably large.

Moreover, in the light of (A.26), (A.141) and (A.148), and recalling that $|u(\check{x})| < 1/2$ and $|u(x_0)| < 1/2$, we have

(A.149) $$|\check{x} - x_0| \leq |(\check{x} - x_0)'| + |(\check{x} - x_0)_N| \leq (1 + 2c_*)q + |(\check{x} - x_0)_N| \leq \check{C}q,$$

where $\check{C}$ is a conveniently large constant as in the statement of the Lemma. By means of (A.148) and (A.144), we conclude that

(A.150) $$(\check{x} - x_0)\cdot \frac{\nabla u(x_0)}{|\nabla u(x_0)|} \leq H_0(u(\check{x})) - H_0(u(x_0)) + \frac{6\,C_\sharp\,q^2}{R}.$$

Let us now consider the set $\Xi$ defined in the statement of Lemma A.1 and take $c_*$ conveniently small: then, thanks to (A.141), (A.140), (A.149), (A.145) and (A.150), we have that
$$\Xi \supseteq \check{\Xi};$$
the proof of (A.22)–(A.23) is thus ended by means of (A.146).

To finish the proof of Lemma A.1, we need now to check the *Lipschitz graph* property of $\Xi_s := \check{\Xi} \cap \{x_{N+1} = s\}$, for any $|s| < 1/2$. The graph property is a straightforward consequence of the *first occurrence* touching point property, hence we focus on the Lipschitz estimate. For this, take $\check{X}, \bar{X} \in \Xi_s$; then,

$$s = u(\bar{x}) = g_{\mathbb{S}(\bar{Y},\check{R})}(\bar{x}) = g_{\mathbb{S}(\check{Y},\check{R})}(\check{x}) = u(\check{x}),$$

for suitable $\check{Y}, \bar{Y} \in \mathbb{R}^{N+1}$, so that $\pi_N \check{Y}, \pi_N \bar{Y} \in \Xi_*$. We will prove that

(A.151) $$|\check{x}_N - \bar{x}_N| \leq \frac{\operatorname{const} q}{R}|(\check{x} - \bar{x})'|,$$

which implies the desired Lipschitz estimate (with constant $\operatorname{const} q/R < 1$). With no loss of generality, we may and do assume that

(A.152) $$\bar{x}_N > \check{x}_N.$$

Let us define

$$\bar{r} := \check{R} + H_{\bar{y}_{N+1},\check{R}}(s) - H_0(\bar{y}_{N+1}),$$

so that

$$\{g_{\mathbb{S}(\bar{Y},\check{R})} = s\} = \partial B_{\bar{r}}(\bar{y}).$$

By (2.57) and (A.143), we have that

(A.153) $$|\bar{x}' - \bar{y}'| \leq \operatorname{const} q,$$

and therefore, from (A.153), we get that

(A.154) $$|\check{x}' - \bar{y}'| \leq |\check{x}' - x_0'| + |x_0' - \bar{x}'| + |\bar{x}' - \bar{y}'| \leq$$
$$\leq \operatorname{const} q,$$

which is less then $\bar{r}$, if $q/R$ is small enough. Thus, by the first occurrence touching property of $g_{\mathbb{S}(\bar{Y},\check{R})}$, $\check{x}$ must be above $\{g_{\mathbb{S}(\bar{Y},\check{R})} = s\}$ (with respect to the $e_N$ direction). This, (A.154) and (A.152) imply that $\check{x}$ is trapped inside a cone with vertex in $\bar{x}$ and slope bounded by $\operatorname{const} q/\bar{r}$, which gives (A.151). This ends the proof of Lemma A.1. □

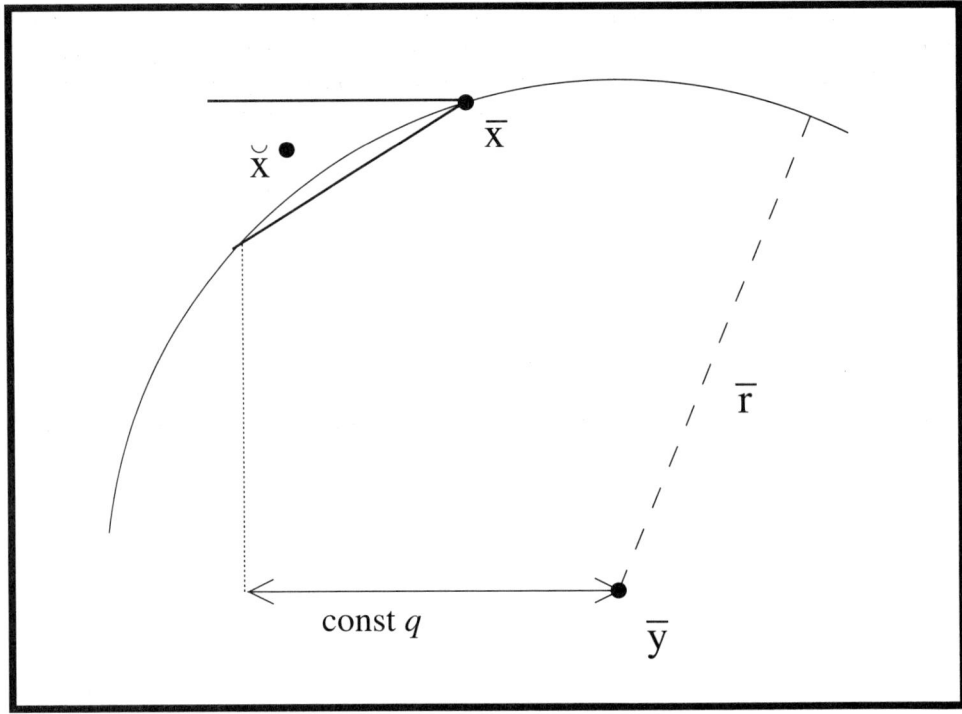

**The cone trapping $\breve{x}$**

By a rotation/translation argument, we deduce from Lemma A.1 the following

LEMMA A.2. *Fix $o \in \mathbb{R}^N$, $\xi \in \mathbb{R}^N$, with $|\xi| = 1$. Let $u$ be a $C^1$-subsolution of (1.5) in*

$$\{x \in \mathbb{R}^N \ s.t. \ |(x-o) \cdot \xi| < l \ and \ |(x-o) - ((x-o) \cdot \xi)\xi| < l\}.$$

*Assume that $\mathbb{S}(Y,R)$ is above $u$ and that $\mathbb{S}(Y,R)$ touches the graph of $u$ at the point $(x_0, u(x_0))$.*

*Suppose that $|u(x_0)| < 1/2$, $|(x_0-o) \cdot \xi| < l/4$, $|(x_0-o) - ((x_0-o) \cdot \xi)\xi| < l/4$. Assume also that*

$$T_{Y,R}x_0 \in \{|(x-o) - ((x-o) \cdot \xi)\xi| = q\} \cap \{(x-o) \cdot \xi = 0\} \qquad and$$

$$y = o - \sqrt{r^2 - q^2}\,\xi \qquad with$$

$$r = r(Y,R) = R - H_0(y_{N+1}) - \frac{\overline{C_0}}{2R}y_{N+1}^2.$$

*Then, there exist universal constants $C_1, C_2 > 1 > c > 0$ such that, if*

$$C_1 \leq q \leq \frac{l}{C_1} \qquad and \qquad 4\sqrt[3]{R} \leq l \leq cR,$$

*the following holds. Let $\Xi$ be the set of points $(\mathfrak{x}, u(\mathfrak{x})) \in \mathbb{R}^N \times \mathbb{R}$ satisfying the following properties:*

- $|(\mathfrak{x}-o) \cdot \xi| < q/15$, $|u(\mathfrak{x})| < 1/2$, $|\mathfrak{x} - x_0| < \check{C}q$;
- *there exists $\hat{Y} \in \mathbb{R}^{N+1}$ such that $\mathbb{S}(\hat{Y}, R/C_2)$ is above $u$ and it touches $u$ at $(\mathfrak{x}, u(\mathfrak{x}))$;*

- $\angle\left(\dfrac{\nabla u(\mathfrak{r})}{|\nabla u(\mathfrak{r})|},\dfrac{\nabla u(x_0)}{|\nabla u(x_0)|}\right) \le \dfrac{C_1 q}{R}$;
- $(\mathfrak{r}-x_0)\cdot \dfrac{\nabla u(x_0)}{|\nabla u(x_0)|} \le \dfrac{C_1 q^2}{R} + H_0(u(\mathfrak{r})) - H_0(u(x_0))$.

Then,
$$\mathfrak{L}^N\left(\pi_\xi(\Xi)\right) \ge c q^{N-1}.$$

More precisely, for any $s \in (-1/2, 1/2)$, there exists a set $\Xi_s \subseteq \Xi \cap \{x_{N+1} = s\}$, which is contained in a Lipschitz graph in the $\xi$-direction, with Lipschitz constant less than 1, and so that, if
$$\check{\Xi} := \bigcup_{s\in(-1/2,1/2)} \Xi_s,$$

we have that
$$\mathfrak{L}^N\left(\pi_\xi(\check{\Xi})\right) \ge c q^{N-1}.$$

With Lemma A.2 in hand, we can now complete the proof of Lemma 4.2, by arguing as follows.

Let us now consider a vector $\xi \in \mathbb{R}^N$ so that $|\xi| = 1$ and
$$\vartheta := \angle(\xi, T_{Y_0,R}x_0 - y_0) = \arcsin\dfrac{q}{r},$$

being $x_0$, $Y_0$ and $R$ the ones in the statement of Lemma 4.2. We also denote $r = r(Y_0, R)$, according to the definition of $r(\cdot,\cdot)$ given in (A.18).

Observe that, by the assumptions of Lemma 4.2 and (2.57),

(A.155)
$$\begin{aligned}\angle(\xi, e_N) &\le \angle(\xi, T_{Y_0,R}x_0 - y_0) + \angle(T_{Y_0,R}x_0 - y_0, e_N) = \\ &= \vartheta + \angle\left(\dfrac{\nabla u(x_0)}{|\nabla u(x_0)|}, e_N\right) \le \\ &\le \mathrm{const}\,\dfrac{q}{R} + \dfrac{\pi}{8} \le \dfrac{\pi}{6}.\end{aligned}$$

Define
$$o := y_0 + \sqrt{r^2 - q^2}\,\xi.$$

We will think $o$ as the origin. Then,

(A.156) $\quad |T_{Y_0,R}x_0 - o| = |y_0 - T_{Y_0,R}x_0|\sin\vartheta = r\cdot\dfrac{q}{r} = q$

and so

(A.157)
$$\begin{aligned}|T_{Y_0,R}x_0 - y_0|^2 &= r^2 = \\ &= |o - y_0|^2 + q^2 = \\ &= |o - y_0|^2 + |T_{Y_0,R}x_0 - o|^2.\end{aligned}$$

Then, (A.157) says that the triangle with vertices in $y_0$, $o$ and $T_{Y_0,R}x_0$ is a right triangle in $o$, that is

(A.158) $\quad (T_{Y_0,R}x_0 - o)\cdot\xi = 0.$

Also, if
$$\check{q} := |(T_{Y_0,R}x_0 - o) - ((T_{Y_0,R}x_0 - o)\cdot\xi)\xi|,$$

then (A.155) and (A.19) give that $q \sim \breve{q}$. Then, the hypotheses of Lemma A.2 being fulfilled (with $\breve{q}$ replacing $q$, possibly scaling $l$ to $\text{const } l$) thanks to (A.155), (A.156) and (A.158). Thence, we deduce that,

(A.159) $$\mathcal{L}^N(\pi_\xi(\Xi)) \geq \text{const } q^{N-1}.$$

Also, in the light of the Lipschitz graph property in Lemma A.2, we have that $\Xi \supseteq \breve{\Xi}$, for an appropriate set $\breve{\Xi}$, with

$$\breve{\Xi} \cap \{x_{N+1} = s\} = \Xi_s = F_s(\Xi_s^\sharp),$$

for any $|s| < 1/2$, for some

$$\Xi_s^\sharp \subseteq \{x \cdot \xi = 0\} \cap \{x_{N+1} = s\}$$

and $|F_s|_{\text{Lip}} \leq 1$. Note that the $\Xi_s^\sharp$'s are all disjoint $(N-1)$-dimensional sets, lying on $\{x_{N+1} = s\}$.

Given $a, b \in \Xi_s^\sharp$, let now $a' := F_s(a)$ and $b' := F_s(b)$. Then, the fact that $F_s$ gives a Lipschitz graph in the $\xi$-direction with $|F_s|_{\text{Lip}} \leq 1$ implies that

$$\angle(a' - b', \xi) \geq \frac{\pi}{4}.$$

Therefore, by means of (A.155),

$$\angle(a' - b', e_N) \geq \frac{\pi}{4} - \frac{\pi}{6} = \frac{\pi}{12}.$$

This says that $\pi_N\big|_{\Xi_s}$ is invertible and that its inverse is a Lipschitz function, with Lipschitz constant bounded by $1/\tan(\pi/12)$. Therefore, if

$$G_s := \pi_\xi \circ \left(\pi_N\big|_{\Xi_s}\right)^{-1},$$

we have that $G_s$ is a Lipschitz function whose range is $\Xi_s^\sharp$, with

$$|G_s|_{\text{Lip}} \leq \frac{1}{\tan(\pi/12)} \leq \text{const}.$$

Hence, using the change of variables formula (see, e.g., page 99 in [**18**]), we deduce that

$$
\begin{aligned}
\operatorname{const} q^{N-1} &\leq \mathcal{L}^N\left(\pi_\xi(\breve{\Xi})\right) = \\
&= \mathcal{L}^N\Big(\bigcup_{s\in(-1/2,1/2)} \Xi_s^\sharp\Big) = \\
&= \int_{-1/2}^{1/2} \mathcal{L}^{N-1}(\Xi_s^\sharp)\, ds = \\
&= \int_{-1/2}^{1/2} \int_{\Xi_s^\sharp} dy\, ds \leq \\
&\leq \int_{-1/2}^{1/2} \int_{G_s^{-1}(\Xi_s^\sharp)} |\det G_s'(x)|\, dx\, ds \leq \\
&\leq \operatorname{const} \int_{-1/2}^{1/2} \int_{G_s^{-1}(\Xi_s^\sharp)} dx\, ds = \\
&= \operatorname{const} \int_{-1/2}^{1/2} \mathcal{L}^{N-1}\bigl(G_s^{-1}(\Xi_s^\sharp)\bigr)\, ds = \\
&= \operatorname{const} \mathcal{L}^N\Big(\bigcup_{s\in(-1/2,1/2)} G_s^{-1}(\Xi_s^\sharp)\Big) = \\
&= \operatorname{const} \mathcal{L}^N\Big(\bigcup_{s\in(-1/2,1/2)} \pi_N\bigl(F_s(\Xi_s^\sharp)\bigr)\Big) = \\
&= \operatorname{const} \mathcal{L}^N\Big(\bigcup_{s\in(-1/2,1/2)} \pi_N(\Xi_s)\Big) \leq \\
&\leq \operatorname{const} \mathcal{L}^N\bigl(\pi_N(\Xi)\bigr),
\end{aligned}
$$

completing the proof of Lemma 4.2.

## A.3. Proof of Lemma 4.3

Let $F_k \subseteq E_k$ be the closed set defined as

$$F_k := \{Z \in L \mid \operatorname{dist}(Z, D_k \cap Q_{l+a}) \leq a\}.$$

If $Q_l \setminus F_k = \emptyset$,

$$\mathcal{L}^N(Q_l \setminus E_k) \leq \mathcal{L}^N(Q_l \setminus F_k) = 0,$$

proving the claim, hence we may and do assume that $Q_l \setminus F_k \neq \emptyset$. Let now $Z \in Q_l \setminus F_k$ and take $Z_* \in F_k$ be so that

(A.160) $$\operatorname{dist}(Z, F_k) = |Z - Z_*| =: r.$$

We use the notation $Z = (z', 0, z_{N+1}) \in L$ and we claim that

(A.161) $$r \leq l + |z'| - \frac{a}{2}.$$

To prove this, we may assume that $r \geq a$, otherwise the claim is proved, and we proceed as follows. First of all, notice that, from (P1) in the statement of

Lemma 4.3, we have that there exists $\check{Z} \in D_k \cap Q_l$. Let $Z_\sharp$ lie on the segment joining $Z$ and $\check{Z}$, at distance $a$ from $\check{Z}$. Then,
$$|Z - Z_\sharp| = |Z - \check{Z}| - a$$
and, therefore, since $\check{Z} \in Q_l$,

$$\begin{aligned}
|Z - Z_\sharp| &\leq \sqrt{(z' - \check{z}')^2 + 1} - a \leq \\
&\leq \sqrt{(|z'| + |\check{z}'|)^2 + 1} - a \leq \\
&\leq \sqrt{(|z'| + l)^2 + 1} - a.
\end{aligned}$$
(A.162)

Also, by construction, $Z_\sharp \in F_k$ and thus

(A.163) $$r \leq |Z - Z_\sharp|.$$

The proof of (A.161) now follows from (A.162) and (A.163) by taking $a > 2$.

Notice now that, since $Z \in Q_l$, (A.161) implies that

(A.164) $$r \leq 2l - \frac{a}{2}.$$

We now claim that

(A.165) $$\mathcal{L}^N\Big(F_{k+1} \cap Q_l \cap B_r(Z)\Big) \geq \bar{c}\, \mathcal{L}^N\Big(Q_l \cap B_r(Z)\Big),$$

for a suitable $\bar{c} \in (0, 1)$, which may depend on the quantity $c$ introduced in (P2) during the statement of Lemma 4.3.

We now begin with the proof of (A.165), which will be completed on page 131.

Since $Z_* \in F_k$, there exists $Z_0 \in D_k \cap Q_{l+a}$ be so that $|Z_* - Z_0| \leq a$. We point out that, in fact,

(A.166) $$|Z_* - Z_0| = a.$$

Indeed, if, by contradiction, $|Z_* - Z_0| < a$, we have that $B_d(Z_*) \cap L \subseteq F_k$ for some $d > 0$, from which it would exists $\hat{Z} \in F_k$ so that $|\hat{Z} - Z| < |Z_* - Z|$, that contradicts the definition of $Z_*$ and proves (A.166).

Also, from (A.160) and (A.166),

(A.167) $$|Z - Z_0| \leq |Z - Z_*| + |Z_* - Z_0| = r + a.$$

We notice that, in fact,

(A.168) $$|Z - Z_0| = r + a.$$

Indeed, if it holded that $|Z - Z_0| < r + a$, take $\tilde{Z} \in \partial B_a(Z_0)$ on the segment joining $Z$ and $Z_0$: then $\tilde{Z} \in F_k$, since $Z_0 \in D_k \cap Q_{l+a}$, and
$$|Z - \tilde{Z}| = |Z - Z_0| - a < r,$$
which contradicts the definition of $r$ and proves (A.168).

Notice that, thanks to (A.167) and (A.168), we have that $Z_*$ belongs to the segment joining $Z$ and $Z_0$.

## A.3. PROOF OF LEMMA 4.3

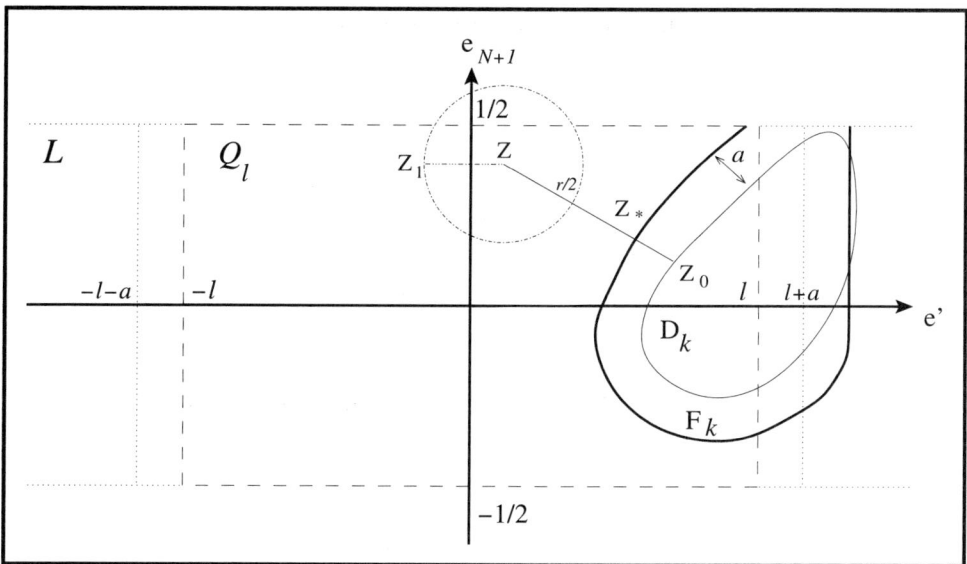

**The proof of Lemma 4.3**

To continue with the proof of (A.165), we need now to distinguish two cases: either $a \leq 2r$ or $a > 2r$. Let us first deal with the case $a \leq 2r$. We claim that there exists $Z_1 \in Q_l$ so that

(A.169) $$|Z - Z_1| = \frac{r}{2} \quad \text{and} \quad B_{r/2}(Z_1) \cap L \subseteq Q_l.$$

To prove (A.169), we need to distinguish two sub-cases. If $z' = 0$, take $Z_1 := Z + re_1/2$. Then,

$$|z_{1,N+1}| = |z_{N+1}| \leq \frac{1}{2}$$

and, recalling (A.164),

$$|z'_1| = \frac{r}{2} < l,$$

showing that $Z_1 \in Q_l$ in this case. Also, exploiting (A.161), if $W = (w', 0, w_{N+1}) \in B_{r/2}(Z_1)$,

$$|w'| \leq r \leq l,$$

proving (A.169) in this sub-case. If, on the other hand, $z' \neq 0$, take

$$Z_1 := Z - \frac{r\,z'}{2\,|z'|};$$

thus,

$$|z_{1,N+1}| = |z_{N+1}| \leq \frac{1}{2}$$

and, since $Z \in Q_l$, recalling also (A.161), we have that

$$\begin{aligned}|z_1'| &= \left|\frac{z'}{|z'|}\left(|z'|-\frac{r}{2}\right)\right| = \\ &= \left||z'|-\frac{r}{2}\right| = \\ &= \max\left\{|z'|-\frac{r}{2},\frac{r}{2}-|z'|\right\} \le \\ &\le \max\left\{l-\frac{r}{2},\frac{l}{2}\right\} < \\ &< l,\end{aligned}$$

which shows that $Z_1 \in Q_l$ in this case.

Let also $W = (w', 0, w_{N+1}) \in B_{r/2}(Z_1) \cap L$; then, we have that

$$\begin{aligned}|w'| &\le |z_1'| + \frac{r}{2} = \\ &= \left|\frac{z'}{|z'|}\left(|z'|-\frac{r}{2}\right)\right| + \frac{r}{2} = \\ &= \left||z'|-\frac{r}{2}\right| + \frac{r}{2} = \\ &= \max\left\{|z'|-\frac{r}{2},\frac{r}{2}-|z'|\right\} + \frac{r}{2} = \\ &= \max\{|z'|, r-|z'|\}\,.\end{aligned}$$

Hence, using that $Z \in Q_l$ and (A.161), we deduce from the above that

$$|w'| \le l\,,$$

proving (A.169) in this sub-case.

This completes the proof of (A.169).

As a consequence of (A.169), we immediately infer from the fact that $B_{r/2}(Z_1) \cap L \subseteq Q_l$ that

$$\hat{W} := Z_1 + \frac{z_1'}{|z_1'|}\frac{r}{2} \in Q_l$$

and therefore

$$l \ge |\hat{w}'| = \left||z_1'| + \frac{r}{2}\right| = |z_1'| + \frac{r}{2}\,,$$

that is

(A.170) $$|z_1'| \le l - \frac{r}{2}\,.$$

We show that this yields that

(A.171) $$|z_1' - z_0'| \le 2l - 2a\,.$$

Indeed, if $r > 2(l-3a)/3$, (A.170) and the fact that $Z_0 \in Q_{l+a}$ imply that

$$|z_1' - z_0'| \le |z_1'| + |z_0'| \le 2l - \frac{r}{2} + a \le \frac{5l}{3} + 2a\,,$$

which proves (A.171) in this case; if, on the other hand, $r \le 2(l-3a)/3$, we have, by (A.168) and (A.169), that

$$|z_1' - z_0'| \le |z_1' - z'| + |z' - z_0'| \le |Z_1 - Z| + |Z - Z_0| = \frac{r}{2} + r + a \le l - 2a$$

in this case. This ends the proof of (A.171).

We now complete the proof of (A.165) in the case $a \leq 2r$, by arguing as follows. We notice that, by construction, and recalling (A.168) and (A.169),

$$\begin{aligned} a + \frac{r}{2} &= |Z - Z_0| - |Z_1 - Z| \leq \\ &\leq |Z_1 - Z_0| \leq \\ &\leq |Z_1 - Z| + |Z - Z_0| = \\ &= \frac{r}{2} + r + a \leq \\ &\leq 5r. \end{aligned} \qquad \text{(A.172)}$$

Furthermore,[8] using (A.171), we have that

$$\begin{aligned} |Z_1 - Z_0| &\leq \sqrt{|z_1' - z_0'|^2 + 1} \leq \\ &\leq \sqrt{(2l - 2a)^2 + 1} \leq \\ &\leq 2l. \end{aligned} \qquad \text{(A.173)}$$

Now notice that, from (A.172),

$$\text{(A.174)} \qquad \mathcal{L}^N\left(D_{k+1} \cap B_{r/2}(Z_1)\right) \geq \mathcal{L}^N\left(D_{k+1} \cap B_{|Z_1-Z_0|/10}(Z_1)\right).$$

Furthermore, by (A.169), $B_{r/2}(Z_1) \cap L$ contains a circular sector of height 1 of a ball of radius $r/2$, while, on the other hand, $B_r(Z) \cap Q_l$ is contained in a circular sector of height 1 of a ball of radius $r$. Therefore,

$$\begin{aligned} \mathcal{L}^N\left(B_{r/2}(Z_1) \cap L\right) &\geq \text{const}\, r^{N-1} \geq \\ &\geq \mathcal{L}^N\left(B_r(Z) \cap Q_l\right). \end{aligned} \qquad \text{(A.175)}$$

Also, by construction, $B_{r/2}(Z_1) \subseteq B_r(Z)$, $D_{k+1} \subseteq F_{k+1} \cap Q_{l+a}$ and $L \cap B_{r/2}(Z_1) \subseteq Q_l$, therefore,

$$\text{(A.176)} \qquad \mathcal{L}^N\left(F_{k+1} \cap Q_l \cap B_r(Z)\right) \geq \mathcal{L}^N\left(D_{k+1} \cap B_{r/2}(Z_1)\right).$$

Finally, from (P2) of Lemma 4.3 (which may be used thanks to (A.173) and (A.172)), we have that

$$\begin{aligned} \mathcal{L}^N\left(D_{k+1} \cap B_{|Z_1-Z_0|/10}(Z_1)\right) &\geq c\,\mathcal{L}^N\left(L \cap B_{|Z_1-Z_0|}(Z_1)\right) \geq \\ &\geq c\,\mathcal{L}^N\left(L \cap B_{r/2}(Z_1)\right), \end{aligned} \qquad \text{(A.177)}$$

where, in the latter estimate, we used again (A.172). Then, (A.165) easily follows in this case from (A.176), (A.174), (A.177) and (A.175).

Let us now deal with the case in which $a > 2r$. In this case,

$$\text{(A.178)} \qquad \frac{r+a}{10} < a.$$

Since $Z \in L$ and $a \geq \text{const} > 0$,

$$\mathcal{L}^N\left(B_{r+a}(Z) \cap L\right) \geq \text{const} > 0.$$

---

[8]The reader will notice that (A.173) is needed in order to use property (P2) of Lemma 4.3 here in the sequel.

In particular, by (P2) of Lemma 4.3, (A.168) and (A.178), we infer from this that

$$\mathcal{L}^N\Big(B_{(r+a)/10}(Z) \cap D_{k+1}\Big) = \mathcal{L}^N\Big(B_{|Z-Z_0|/10}(Z) \cap D_{k+1}\Big) \geq$$
$$\geq \text{const } \mathcal{L}^N\Big(B_{|Z-Z_0|}(Z) \cap L\Big) =$$
$$= \text{const } \mathcal{L}^N\Big(B_{r+a}(Z) \cap L\Big) \geq$$
$$\geq \text{const} > 0$$

thence, there exists
$$Z_\sharp \in D_{k+1} \cap B_{(r+a)/10}(Z).$$

Also, the fact that $Z \in Q_l$ and (A.178) give that
$$D_{k+1} \cap B_{(r+a)/10}(Z) \subseteq Q_{l+a},$$

thence $Z_\sharp \in Q_{l+a}$. Hence, the fact that $Z_\sharp \in B_{(r+a)/10}(Z)$ implies that

(A.179) $$Q_l \cap B_r(Z) \subseteq Q_l \cap B_a(Z_\sharp),$$

while the fact that $Z_\sharp \in D_{k+1}$ implies that

(A.180) $$Q_l \cap B_a(Z_\sharp) \subseteq F_{k+1}.$$

Then, from (A.179) and (A.180),
$$Q_l \cap B_r(Z) \subseteq F_{k+1}$$

and, therefore,
$$F_{k+1} \cap Q_l \cap B_r(Z) = Q_l \cap B_r(Z).$$

This proves (A.165) in this case (with $\bar{c} = 1$).

Having completed the proof of (A.165) we now take a finite overlapping cover $\mathfrak{C}$ of $Q_l \setminus F_k$ with balls of radius $r$, in order to end the proof of Lemma 4.3. Thus, using such cover,

$$\mathcal{L}^N\Big(F_{k+1} \cap (Q_l \setminus F_k)\Big) = \mathcal{L}^N\Big(F_{k+1} \cap Q_l \cap (Q_l \setminus F_k)\Big) \geq$$
$$\geq \text{const} \sum_{B_r(Z) \in \mathfrak{C}} \mathcal{L}^N\Big(F_{k+1} \cap Q_l \cap B_r(Z)\Big).$$

Then, using (A.165), we deduce from the above that

$$\text{const } \bar{c}^{-1} \mathcal{L}^N\Big(F_{k+1} \cap (Q_l \setminus F_k)\Big) \geq \sum_{B_r(Z) \in \mathfrak{C}} \mathcal{L}^N\Big(Q_l \cap B_r(Z)\Big) \geq$$
$$\geq \mathcal{L}^N\Big(Q_l \cap \Big(\bigcup_{B_r(Z) \in \mathfrak{C}} B_r(Z)\Big)\Big) \geq$$
$$\geq \mathcal{L}^N\Big(Q_l \cap (Q_l \setminus F_k)\Big) =$$
(A.181) $$= \mathcal{L}^N(Q_l \setminus F_k).$$

Furthermore, since $D_k \subseteq D_{k+1}$, we have that $F_k \subseteq F_{k+1}$ and so

(A.182) $$Q_l \setminus F_{k+1} \subseteq \Big(Q_l \setminus F_k\Big) \setminus \Big(F_{k+1} \cap (Q_l \setminus F_k)\Big).$$

## A.3. PROOF OF LEMMA 4.3

Hence, by using (A.182) and (A.181),

$$\begin{aligned}\mathfrak{L}^N(Q_l \setminus F_{k+1}) &\leq \mathfrak{L}^N(Q_l \setminus F_k) - \mathfrak{L}^N\Big(F_{k+1} \cap (Q_l \setminus F_k)\Big) \leq \\ &\leq (1-\hat{c})\,\mathfrak{L}^N(Q_l \setminus F_k)\,,\end{aligned}$$

for a suitable $\hat{c}$. Therefore, iterating the above estimate,

$$\mathfrak{L}^N(Q_l \setminus F_k) \leq (1-\hat{c})^k\,\mathfrak{L}^N(Q_l)\,.$$

This completes the proof of Lemma 4.3 since $E_k \supseteq F_k$ by construction.

# APPENDIX B

# Summary of elementary lemmata

We collect here some lemmata that are in use during the proofs of the main results. We will skip the proofs of most of these lemmata, since they are quite elementary (a detailed proof of them, however, may be found in [**30**]).

LEMMA B.1. *For any $0 \leq s \leq t \leq \theta^*$,*
$$h_0(-1+t) - h_0(-1+s) \geq c(t^p - s^p),$$
*for a suitable universal constant $c > 0$.*

LEMMA B.2. *Let $U$ be an open subset of $\mathbb{R}$. Let $g \in C^2(U)$ and assume that $g$ has no critical points. Define*

(B.1) $$\Psi^{y,l}(x) := g(|x-y| - l)$$

*Then, for $t = |x-y| - l \in U$ and $x \neq y$, we have*

(B.2) $$\Delta_p(\Psi^{y,l}(x)) = (p-1)g''(t)g'^{(p-2)}(t) + g'^{(p-1)}(t)\frac{N-1}{|x-y|}$$

LEMMA B.3. *Let $I \ni 0$ be an interval of $\mathbb{R}$ and let $h \in C^1(I)$ satisfy $h(s) > 0$ for any $s \in I$. Let*
$$H(s) := \int_0^s \frac{(p-1)^{1/p} \, d\zeta}{(p\,h(\zeta))^{1/p}}, \qquad \forall s \in I.$$
*Define also $g$ as the inverse of $H$, that is $g(t) := H^{-1}(t)$ for any $t \in H(I)$. Then, $g \in C^2(H(I))$ and*
$$g'(t) = \left(\frac{p}{p-1} h(g(t))\right)^{1/p}$$
$$g''(t) = \frac{(p\,h(g(t)))^{(2-p)/p}}{(p-1)^{2/p}} h'(g(t)),$$
*for any $t \in H(I)$.*

We recall now the maximum and comparison principles needed for our purposes. First of all, in [**9**] (see in particular Theorem 1.4 there) the following result was obtained:

THEOREM B.4 (Strong Comparison Principle I). *Let $\Omega$ be an open (not necessarily bounded nor connected) subset of $\mathbb{R}^N$, $\Lambda \in \mathbb{R}$ and $u, v \in C^1(\Omega)$ satisfy*

(B.3) $$-\Delta_p u + \Lambda u \leq -\Delta_p(v) + \Lambda v, \qquad u \leq v \text{ in } \Omega.$$

*Define $Z_{u,v} = \{x \in \Omega : |Du(x)| + |Dv(x)| = 0\}$ if $p \neq 2$, $Z_{u,v} = \emptyset$ if $p = 2$. If $x_0 \in \Omega \setminus Z_{u,v}$ and $u(x_0) = v(x_0)$, then $u = v$ in the connected component of $\Omega \setminus Z_{u,v}$ containing $x_0$.*

An easy consequence of the above result is the following one (see §3 in [**30**] for further details):

COROLLARY B.5 (Strong Comparison Principle II). *Let $\Omega$ be an open (not necessarily bounded nor connected) subset of $\mathbb{R}^N$, and $u, v \in C^1(\Omega)$ satisfy*

(B.4) $\qquad -\Delta_p u + f(u) \leq -\Delta_p(v) + f(v), \qquad u \leq v \text{ in } \Omega,$

*with $f$ locally Lipschitz continuous. Define $Z_{u,v} = \{x \in \Omega : |Du(x)| + |Dv(x)| = 0\}$ if $p \neq 2$, $Z_{u,v} = \emptyset$ if $p = 2$. If $x_0 \in \Omega \setminus Z_{u,v}$ and $u(x_0) = v(x_0)$, then $u = v$ in the connected component of $\Omega \setminus Z_{u,v}$ containing $x_0$.*

As well known, the "dangerous" points in dealing with $p$-Laplace operators are the ones in which the gradient vanishes, due to lack of ellipticity. Next result, proved in [**37**] (see also [**30**] for details), will help us in dealing with this circumstance.

THEOREM B.6 (Strong Maximum Principle and Hopf's Lemma). *Let $\Omega$ be an open connected (not necessarily) bounded set in $\mathbb{R}^N$ and suppose that $u \in C^1(\Omega)$, $u \geq 0$ in $\Omega$, weakly solves*

$$-\Delta_p u + cu^q = g \geq 0 \quad \text{in} \quad \Omega$$

*with $q \geq p-1$, $c \geq 0$ and $g \in L^\infty_{loc}(\Omega)$. If $u$ is not identically zero, then $u > 0$ in $\Omega$. Moreover, for any point $x_0 \in \partial\Omega$ where the interior sphere condition is satisfied, and such that $u$ is $C^1$ in a neighborhood of $\Omega \cup \{x_0\}$ and $u(x_0) = 0$, we have that $\frac{\partial u}{\partial s} > 0$ for any inward directional derivative.*

Following are some easy result on the geometry of Euclidean spheres. Though elementary, we give full details of their proofs, in order to take care of the constants involved.

LEMMA B.7. *Let $r > q \geq 0$. Fix $\epsilon > 0$ and let $c_1$ and $c_2$ be non-negative and so that*

(B.5) $\qquad \max\{c_1, c_2\} \leq \min\{\sqrt{\epsilon}/3,\, 1/2\}.$

*Suppose[1] that $z \in B_r(y) \subset \mathbb{R}^N$, with $|z' - y'| \leq c_1 q$. Then,*

(B.6) $$\partial B_r(y) \cap \left\{|x' - z'| \leq c_2 q\right\} \cap \left\{x_N \geq y_N\right\} \subseteq$$
$$\subseteq \left\{x_N \geq z_N - \frac{\epsilon q^2}{r}\right\}.$$

PROOF. Take $w \in \partial B_r(y) \cap \left\{|x' - z'| \leq c_2 q\right\} \cap \left\{x_N \geq y_N\right\}$. For any $x \in \mathbb{R}^N$, let

$$\hat{x} := \frac{x - y}{r}.$$

Then, by construction,

$$\hat{z} \in B_1(0),$$
$$|\hat{z}'| \leq \frac{c_1 q}{r} \qquad \text{and}$$
$$\hat{w} \in \partial B_1(0) \cap \left\{|\hat{x}' - \hat{z}'| \leq \frac{c_2 q}{r}\right\} \cap \left\{\hat{x}_N \geq 0\right\}.$$

Let also $t \geq 0$ so that

$$\hat{b} := \hat{z} + t e_N \in \partial B_1(0).$$

---

[1] Notation remark: in Lemma B.7, for definiteness, the balls are assumed to be closed.

We claim that

(B.7) $$\hat{w}_N \geq \hat{b}_N - \frac{\epsilon q^2}{r^2}.$$

To prove this, first note that, if $\hat{w}_N \geq \hat{b}_N$, (B.7) is obvious; hence, we may assume that

(B.8) $$\hat{w}_N < \hat{b}_N.$$

Also, if $\hat{b}_N \leq 0$, (B.7) would follow from the fact that $\hat{w}_N \geq 0$, thus we may also suppose that

(B.9) $$\hat{b}_N > 0.$$

Also,

$$\begin{aligned}
1 &= |\hat{b}'|^2 + |\hat{b}_N|^2 = \\
&= |\hat{z}'|^2 + |\hat{b}_N|^2 \leq \\
&\leq \frac{c_1^2 q^2}{r^2} + |\hat{b}_N|^2 \leq \\
&\leq c_1^2 + |\hat{b}_N|^2,
\end{aligned}$$

which, together with (B.9), implies that

(B.10) $$\hat{b}_N \geq \frac{1}{2},$$

thanks to (B.5). Furthermore,

$$\begin{aligned}
|\hat{z}'|^2 + |\hat{b}_N|^2 &= |\hat{b}'|^2 + |\hat{b}_N|^2 = \\
&= |\hat{b}|^2 = \\
&= 1 = \\
&= |\hat{w}|^2 = \\
&= |\hat{w}'|^2 + |\hat{w}_N|^2 \leq \\
&\leq \left(|\hat{z}'| + |\hat{w}' - \hat{z}'|\right)^2 + |w_N|^2,
\end{aligned}$$

and so, by (B.5)

$$\begin{aligned}
|\hat{b}_N|^2 - |\hat{w}_N|^2 &\leq |\hat{w}' - \hat{z}'|^2 + 2|\hat{z}'||\hat{w}' - \hat{z}'| \leq \\
&\leq (c_2^2 + 2c_1 c_2)\frac{q^2}{r^2} \leq \\
&\leq \frac{\epsilon q^2}{2r^2}.
\end{aligned}$$

From this, the fact that $\hat{w}_N \geq 0$, (B.8) and (B.10), we conclude that

$$\frac{\epsilon q^2}{2r^2} \geq (\hat{b}_N + \hat{w}_N)(\hat{b}_N - \hat{w}_N) \geq \frac{1}{2}(\hat{b}_N - \hat{w}_N),$$

which proves (B.7).

By using (B.7), we gather that
$$\begin{aligned} w_N - z_N &= r(\hat{w}_N - \hat{z}_N) = \\ &= r(\hat{w}_N - \hat{b}_N + t) \geq \\ &\geq r(\hat{w}_N - \hat{b}_N) \geq -\frac{\epsilon q^2}{r}, \end{aligned}$$
which gives (B.6). □

COROLLARY B.8. *Let $r > q \geq 0$ and fix*

(B.11) $$\kappa \in \left(0, \frac{r}{10q}\right].$$

*Let us suppose that $v = (v', \kappa q^2/r) \in \mathbb{R}^N$ is above $B_r(y)$ with respect to the $e_N$ direction. Let us assume also that $|v' - y'| \leq cq$, with*

(B.12) $$c \leq \min\left\{\frac{\sqrt{\kappa}}{3}, \frac{1}{2}\right\}.$$

*Then,*
$$\partial B_r(y) \cap \{|x' - y'| \geq 4\kappa q\} \subseteq \{x_N < 0\}.$$

PROOF. Take $w \in \partial B_r(y) \cap \{|x' - y'| \geq 4\kappa q\}$. Let also $t \geq 0$ be so that

(B.13) $$p := v - te_N \in \partial B_r(y) \text{ with } p_N \geq y_N.$$

Note that

(B.14) $$|p' - y'| = |v' - y'| \leq cq.$$

Also, by our assumptions,
$$\begin{aligned} w_N &\leq y_N + |w_N - y_N| = \\ &= y_N + \sqrt{r^2 - |w' - y'|^2} \leq \\ &\leq y_N + \sqrt{r^2 - 16\kappa^2 q^2}. \end{aligned}$$
(B.15)

Let $z := y + re_N$. We now apply Lemma B.7 with $\epsilon := \kappa$ and $c_1 := c_2 := c$, recalling (B.12). Indeed, by (B.13), (B.14) and Lemma B.7,
$$\begin{aligned} p \in \partial B_r(y) \cap \{|x' - z'| \leq c_2 q\} \cap \{x_N \geq y_N\} &\subseteq \\ \subseteq \left\{x_N \geq z_N - \frac{\kappa q^2}{r}\right\} &= \\ = \left\{x_N \geq y_N + r - \frac{\kappa q^2}{r}\right\}. \end{aligned}$$
(B.16)

Exploiting (B.15) and (B.16), we get
$$\begin{aligned} w_N &\leq p_N - r + \frac{\kappa q^2}{r} + \sqrt{r^2 - 16\kappa^2 q^2} \leq \\ &\leq v_N - r + \frac{\kappa q^2}{r} + \sqrt{r^2 - 16\kappa^2 q^2} = \\ &= \frac{2\kappa q^2}{r} - r + \sqrt{r^2 - 16\kappa^2 q^2} \leq \\ &\leq \frac{2\kappa q^2}{r} - r + r - \frac{8\kappa^2 q^2}{r} < 0, \end{aligned}$$

which is the desired result. □

COROLLARY B.9. *Let $a > 0$, $r > q \geq 0$ so that*

(B.17) $$\frac{q}{r} \leq \frac{K}{a^2}$$

*and*

(B.18) $$\frac{q}{r} \leq \frac{a}{8K}.$$

*Let us suppose that $v = (v', Kq^2/r) \in \mathbb{R}^N$ is above $B_r(y)$ with respect to the $e_N$ direction. Let us assume also that $|v' - y'| \leq \hat{c}q$, with*

(B.19) $$\hat{c} \leq \min\left\{\frac{\sqrt{K}}{3}, \frac{2K}{a}\right\}.$$

*Then,*

(B.20) $$\partial B_r(y) \cap \left\{|x' - y'| \geq aq\right\} \subseteq \{x_N < 0\}.$$

PROOF. Let $\bar{q} := 4Kq/a$, $\kappa := a^2/(16K)$, $c := a\hat{c}/(4K)$. Note that $\bar{q} < r$ due to (B.18). What is more, $v = (v', \kappa \bar{q}^2/r)$ and $|v' - y'| \leq c\bar{q}$. Also, (B.11) and (B.12) are fulfilled thanks to (B.17) and (B.19). Thus, by Corollary B.8 (applied with $\bar{q}$ instead of $q$),
$$\partial B_r(y) \cap \left\{|x' - y'| \geq 4\kappa \bar{q}\right\} \subseteq \{x_N < 0\},$$
which is (B.20). □

We now point out some observations on rotation hypersurfaces in $\mathbb{R}^{N+1}$. First, the normal of a rotation surface is in the space[2] generated by the radial direction and $e_{N+1}$, as showed by the next result:

LEMMA B.10. *Fix $y \in \mathbb{R}^N$. Let $f \in C^1(\mathbb{R}, \mathbb{R})$ and define*
$$\Phi(x) := f(|x - y|).$$
*Let $\nu(x)$ be a normal vector to the surface $\{x_{N+1} = \Phi(x)\}$ at the point $(x, \Phi(x))$. Then, if $x \neq y$, $\nu(x)$ belongs to the space spanned by $x - y$ and $e_{N+1}$. If $x = y$ and $f'(0) = 0$, the same result holds.*

PROOF. Assume $x \neq y$. By construction,
$$\nu(x) = \alpha\Big(\nabla\Phi(x), -1\Big),$$
for some $\alpha \in \mathbb{R}$. Therefore,
$$\nu(x) = \frac{\alpha f'(|x-y|)}{|x-y|}(x-y) - \alpha e_{N+1},$$
thus proving the claim if $x \neq y$.

If, on the other hand, $x = y$ and $f'(0) = 0$, then $\nu(x) = \alpha e_{N+1}$, for some $\alpha \in \mathbb{R}$, hence completing the proof of the claim. □

The next result will relate the "center" of a rotation hypersurface with the normal at any point:

---

[2]And, in fact, this property characterizes the rotation surfaces, as pointed out to us by Rajko Quarta Marcon, an undergraduate in Tor Vergata.

LEMMA B.11. *Fix $y \in \mathbb{R}^N$. Let $f \in C^1(\mathbb{R}, \mathbb{R})$ and define*
$$\Phi(x) := f(|x - y|).$$
*Let us define the following hypersurface in $\mathbb{R}^{N+1}$:*
$$\Sigma := \left\{ \big(x, \Phi(x)\big) \,\Big|\, x \in \mathbb{R}^N \right\}.$$
*Let us consider the normal $\nu(x)$ at a point $\big(x, \Phi(x)\big) \in \Sigma$, given by*
$$\nu(x) = \big(\nu_1(x), \ldots, \nu_{N+1}(x)\big) := \frac{\big(-\nabla\Phi(x),\, 1\big)}{\sqrt{1 + |\nabla\Phi(x)|^2}}.$$
*Then, for any $x \in \mathbb{R}^N \setminus \{y\}$, the vectors*
$$x - y \quad \text{and} \quad \big(\nu_1(x), \ldots, \nu_N(x)\big)$$
*are parallel.*

PROOF. If $(\nu_1(x), \ldots, \nu_N(x)) = 0$, there is nothing to prove, so we may assume $(\nu_1(x), \ldots, \nu_N(x)) \neq 0$. For this reason,
$$\nabla\Phi(x) = f'\big(|x - y|\big) \frac{x - y}{|x - y|},$$
thus
$$f'\big(|x - y|\big) \neq 0.$$
Let
$$a(x) := \frac{|x - y|\sqrt{1 + |\nabla\Phi(x)|^2}}{f'\big(|x - y|\big)} \in \mathbb{R}.$$
Then,
$$a(x)\big(\nu_1(x), \ldots, \nu_N(x)\big) = x - y,$$
proving the claim. □

Next result is an explicit computation on the differential of the unit normal of a hypersurface (up to a sign, such quantity is sometimes referred to as Second Fundamental Form or Shape Operator):

LEMMA B.12. *Let $\Psi \in C^1(\mathbb{R}^N, \mathbb{R})$. Let $\Sigma$ be the hypersurface defined by*
$$\Sigma := \left\{ \big(x, \Psi(x)\big) \,\Big|\, x \in \mathbb{R}^N \right\}.$$
*Let $X = X(x) := (x, \Psi(x))$ and consider the unit normal to $\Sigma$ at the point $X$, given by*
$$\hat{\nu}(x) := \frac{\big(-\nabla\Psi(x),\, 1\big)}{\sqrt{1 + |\nabla\Psi(x)|^2}} \in S^N.$$
*For $X = (x, \Psi(x))$, let also*
$$\nu(X) = \nu(x, \Psi(x)) := \hat{\nu}(x).$$

Let[3] $D_X\nu : T_X\Sigma \longrightarrow \mathbb{R}^{N+1}$ be the differential map. Then, for any $W = (w, w_{N+1}) \in T_X\Sigma$,

(B.21) $\quad D_X\nu(X)[W] =$

$$= \begin{pmatrix} \dfrac{-(1+|\nabla\Psi(x)|^2)\partial_{1j}\Psi(X)w_j + \partial_1\Psi(X)\,\partial_k\Psi(X)\,\partial_{kj}\Psi(X)\,w_j}{(1+|\nabla\Psi(x)|^2)^{3/2}} \\ \vdots \\ \dfrac{-(1+|\nabla\Psi(x)|^2)\partial_{Nj}\Psi(X)w_j + \partial_N\Psi(X)\,\partial_k\Psi(X)\,\partial_{kj}\Psi(X)\,w_j}{(1+|\nabla\Psi(x)|^2)^{3/2}} \\ \dfrac{-\partial_k\Psi(X)\,\partial_{kj}\Psi(X)\,w_j}{(1+|\nabla\Psi(x)|^2)^{3/2}} \end{pmatrix}.$$

PROOF. Let $W \in T_X\Sigma$. Then, for some $v \in \mathbb{R}^N$,

$$W = \frac{d}{dt}\Big(x+tv,\,\Psi(x+tv)\Big)\Big|_{t=0} = \Big(v,\,\nabla\Psi(x)\cdot v\Big),$$

that is $w_{N+1} = \nabla\Psi(x) \cdot w$, or, equivalently

(B.22) $\quad T_X\Sigma = \{(w, \nabla\Psi(x)\cdot w) \mid w \in \mathbb{R}^N\}.$

Because of this,

$$\begin{aligned} D_X\nu(X)[W] &= \frac{d}{dt}\nu\Big(x+tw,\,\Psi(x+tw)\Big)\Big|_{t=0} = \\ &= \frac{d}{dt}\hat{\nu}(x+tw)\Big|_{t=0}, \end{aligned}$$

from which a straightforward calculation gives the claim. $\square$

REMARK B.13. In relation with Lemma B.12 above, we notice that, since $\nu(\Sigma) \subseteq S^N$ and $T_{\nu(X)}S^N = T_X\Sigma$, we may think $D_X\nu$ as a linear map from $T_X\Sigma$ to itself. Using (B.21) and (B.22), one deduces that $D_X\nu : T_X\Sigma \longrightarrow T_X\Sigma$ may thus be represented in matrix form as

(B.23) $\quad \Big(D_X\nu\Big)_{ij} = \dfrac{-(1+|\nabla\Psi|^2)\partial_{ij}\Psi + \partial_i\Psi\partial_k\Psi\partial_{kj}\Psi}{(1+|\nabla\Psi|^2)^{3/2}}$

for $1 \leq i, j \leq N$ (where, of course, the summation over the index $k$ is understood here above).

With this, we now point out an explicit computation of the curvatures of the rotation surfaces:

LEMMA B.14. Let $\Phi \in C^2((0,+\infty), \mathbb{R})$ and

$$\Sigma := \Big\{\big(x,\,\Phi(|x|)\big) \mid x \in \mathbb{R}^N\Big\}.$$

---

[3] As standard, given a manifold $M$ and a point $X \in M$, we denote by $T_XM$ the tangent space at $X$. Also, as usual, $S^N := \{X \in \mathbb{R}^{N+1} \mid |X| = 1\}$.

Then, the principal curvatures[4] of $\Sigma$ are given by

$$\kappa_1 = \ldots = \kappa_{N-1} = \frac{\Phi'}{|x|\sqrt{1+(\Phi')^2}}$$

$$\kappa_N = \frac{\Phi''}{(1+|\Phi'|^2)^{3/2}}.$$

PROOF. We set $\Psi(x) := \Phi(|x|)$ and, after some easy computation, we infer from (B.23) that

(B.24)
$$\left(D_X \nu\right)_{ij} = (1+(\Phi')^2)^{-3/2}\left[-\frac{(1+(\Phi')^2)\Phi'\delta_{ij}}{|x|} + \frac{(1+(\Phi')^2)\Phi' - |x|\Phi''}{|x|^3} x_i x_j\right].$$

Notice now that, up to rotation, we may assume that the point $X = (x, \Phi(|x|))$, in which we compute the principal curvatures, is of the form $x = |x|\, e_N$; hence, from (B.24)

$$\left(D_X \nu\right)_{ij} = (1+(\Phi')^2)^{-3/2}\left[-\frac{(1+(\Phi')^2)\Phi'\delta_{ij}}{|x|} + \frac{(1+(\Phi')^2)\Phi' - |x|\Phi''}{|x|} \delta_{iN}\delta_{jN}\right],$$

and so the desired claim easily follows. □

---

[4]As standard, the principal curvature of a surface $\Sigma$ at the point $\bar{X}$ are here defined as the eigenvalues of $-D_X\nu(\bar{X}) : T_{\bar{X}}\Sigma \longrightarrow T_{\bar{X}}\Sigma$.

# Bibliography

[1] G. Alberti, L. Ambrosio, X. Cabré, *On a long-standing conjecture of E. De Giorgi: symmetry in 3D for general nonlinearities and a local minimality property*, Acta Appl. Math. 65 (2001), no. 1-3, 9–33.

[2] L. Ambrosio, X. Cabré, *Entire solutions of semilinear elliptic equations in $\mathbb{R}^3$ and a conjecture of De Giorgi*, J. Amer. Math. Soc. 13 (2000), no. 4, 725–739.

[3] S. S. Antman, *Nonuniqueness of equilibrium states for bars in tension*, J. Math. Anal. Appl. (1973), 333–349.

[4] M. T. Barlow, R. F. Bass, C. Gui, *The Liouville property and a conjecture of De Giorgi*, Comm. Pure Appl. Math. 53 (2000), no. 8, 1007–1038.

[5] H. Berestycki, L. A. Caffarelli, L. Nirenberg, *Further qualitative properties for elliptic equations in unbounded domains*, Ann. Sc. Norm. Super. Pisa, Cl. Sci. (4) 25 (1997), no. 1-2, 69–94.

[6] H. Berestycki, F. Hamel, R. Monneau, *One-dimensional symmetry of bounded entire solutions of some elliptic equations*, Duke Math. J. 103 (2000), no. 3, 375–396.

[7] G. Bouchitté, *Singular perturbations of variational problems arising from a two-phase transition model*, Appl. Math. Optim. 21 (1990), no. 3, 289–314.

[8] L.A. Caffarelli, X. Cabré, *Fully nonlinear elliptic equations*, American Mathematical Society Colloquium Publications, 43. American Mathematical Society, Providence, RI, 1995.

[9] L. Damascelli, *Comparison theorems for some quasilinear degenerate elliptic operators and applications to symmetry and monotonicity results*, Ann. Inst. H. Poincaré. Analyse non linéaire, 15(4), 1998, 493–516.

[10] L. Damascelli and B. Sciunzi, *Harnack Inequalities, Maximum and Comparison Principles, and Regularity of positive solutions of m-Laplace equations*, to appear in Calc. Var. Partial Differential Equations,

[11] D. Danielli, N. Garofalo, *Properties of entire solutions of non-uniformly elliptic equations arising in geometry and in phase transitions*, Calc. Var. Partial Differential Equations 15 (2002), no. 4, 451–491.

[12] A. Farina, *One-dimensional symmetry for solutions of quasilinear equations in $\mathbb{R}^2$*, Bollettino UMI 8 6-B (2003), 685–692.

[13] A. Farina, *Rigidity and one-dimensional symmetry for semilinear elliptic equations in the whole of $\mathbb{R}^N$ and in half spaces*, Adv. Math. Sci. Appl. 13 (2003), no. 1, 65–82.

[14] E. De Giorgi, *Convergence problems for functionals and operators*, Proceedings of the International Meeting on Recent Methods in Nonlinear Analysis (Rome, 1978), 131–188, Pitagora, Bologna, 1979.

[15] E. DiBenedetto, $C^{1+\alpha}$ *local regularity of weak solutions of degenerate elliptic equations*, Nonlinear Anal. 7 (1983), no. 8, 827–850.

[16] J. Dolbeault, R. Monneau, *On a Liouville type theorem for isotropic homogeneous fully nonlinear elliptic equations in dimension two*, Ann. Sc. Norm. Super. Pisa Cl. Sci. (5) 2 (2003), no. 1, 181–197.

[17] Y. Du, L. Ma, *Some Remarks on De Giorgi conjecture*, Proc. Amer. Math. Soc., 131(2002), 2415–2422.

[18] L.C. Evans, R.F. Gariepy, *Measure theory and fine properties of functions*, Studies in Advanced Mathematics. CRC Press, Boca Raton, FL, 1992.

[19] D. Gilbarg, N.S. Trudinger, *Elliptic partial differential equations of second order*, Classics in Mathematics. Springer-Verlag, Berlin, 2001.

[20] E. Giusti, *Minimal surfaces and functions of bounded variation*, Monographs in Mathematics, 80. Birkhäuser Verlag, Basel, 1984.

[21] N.A. Ghoussoub, C. Gui, *On a conjecture of De Giorgi and some related problems*, Math. Ann. 311 (1998), no. 3, 481–491.

[22] N. A. Ghoussoub, C. Gui, *On the De Giorgi conjecture in dimensions 4 and 5*, Ann. Math. 157 (2003), no. 1, 313–334.

[23] M.E. Gurtin, *On a theory of phase transitions with interfacial energy*, Arch. Rational Mech. Anal. (1985), no. 3, 187–212.

[24] D. Jerison, R. Monneau, *The existence of a symmetric global minimizer on $\mathbb{R}^{n-1}$ implies the existence of a counter-example to a conjecture of De Giorgi in $\mathbb{R}^n$* , C. R. Acad. Sci. Paris 333 (2001), no. 5, 427–431.

[25] L. Modica, S. Mortola, *Un esempio di $\Gamma^-$-convergenza*, Boll. Un. Mat. Ital. B (5) 14 (1977), no. 1, 285–299.

[26] L. Modica, *The gradient theory of phase transitions and the minimal interface criterion*, Arch. Rational Mech. Anal. 98 (1987), no. 2, 123–142.

[27] A. Petrosyan, E. Valdinoci, *Geometric properties of Bernoulli-type minimizers*, Interfaces Free Bound. 7 (2005), 55–78.

[28] A. Petrosyan, E. Valdinoci, *Density estimates for a degenerate/singular phase-transition model*, SIAM J. Math. Anal. 36 (2005), no. 4, 1057–1079.

[29] J. S. Rowlinson, *Translation of J. D. van der Waals' "The thermodynamic theory of capillarity under the hypothesis of a continuous variation of density"*, J. Statist. Phys. (1979), no. 2, 197–244.

[30] B. Sciunzi, E. Valdinoci, *Mean curvature properties for p-Laplace phase transitions*, J. Eur. Math. Soc. (JEMS) 7 (2005), no. 3, 319–359.

[31] V. O. Savin, *Phase Transitions: Regularity of Flat Level Sets*, Ph.D. thesis at the University of Texas at Austin, 2003.

[32] V. O. Savin, manuscript, 2004.

[33] P. Sternberg, *The effect of a singular perturbation on nonconvex variational problems*, Arch. Ration. Mech. Anal. 101 (1988), no. 3, 209–260.

[34] P. Tolksdorf, *Regularity for a more general class of quasilinear elliptic equations*, J. Differential Equations 51 (1984), no. 1, 126–150.

[35] P. Tolksdorf, *On the dirichlet problem for quasilinear equations in domains with conical boundary points*, Comm. Partial Differential Equations (1983), no. 7, 773–817.

[36] E. Valdinoci, *Plane-like minimizers in periodic media: jet flows and Ginzburg-Landau-type functionals*, J. Reine Angew. Math 574 (2004), 147–185.

[37] J. L. Vazquez, *A strong maximum principle for some quasilinear elliptic equations*, Appl. Math. Optim., 1984, 191 – 202.

★ ★ ★

## Editorial Information

To be published in the *Memoirs*, a paper must be correct, new, nontrivial, and significant. Further, it must be well written and of interest to a substantial number of mathematicians. Piecemeal results, such as an inconclusive step toward an unproved major theorem or a minor variation on a known result, are in general not acceptable for publication. Papers appearing in *Memoirs* are generally at least 80 and not more than 200 published pages in length. Papers less than 80 or more than 200 published pages require the approval of the Managing Editor of the Transactions/Memoirs Editorial Board.

As of March 31, 2006, the backlog for this journal was approximately 13 volumes. This estimate is the result of dividing the number of manuscripts for this journal in the Providence office that have not yet gone to the printer on the above date by the average number of monographs per volume over the previous twelve months, reduced by the number of volumes published in four months (the time necessary for preparing a volume for the printer). (There are 6 volumes per year, each containing at least 4 numbers.)

A Consent to Publish and Copyright Agreement is required before a paper will be published in the *Memoirs*. After a paper is accepted for publication, the Providence office will send a Consent to Publish and Copyright Agreement to all authors of the paper. By submitting a paper to the *Memoirs*, authors certify that the results have not been submitted to nor are they under consideration for publication by another journal, conference proceedings, or similar publication.

## Information for Authors

*Memoirs* are printed from camera copy fully prepared by the author. This means that the finished book will look exactly like the copy submitted.

The paper must contain a *descriptive title* and an *abstract* that summarizes the article in language suitable for workers in the general field (algebra, analysis, etc.). The *descriptive title* should be short, but informative; useless or vague phrases such as "some remarks about" or "concerning" should be avoided. The *abstract* should be at least one complete sentence, and at most 300 words. Included with the footnotes to the paper should be the 2000 *Mathematics Subject Classification* representing the primary and secondary subjects of the article. The classifications are accessible from www.ams.org/msc/. The list of classifications is also available in print starting with the 1999 annual index of *Mathematical Reviews*. The Mathematics Subject Classification footnote may be followed by a list of *key words and phrases* describing the subject matter of the article and taken from it. Journal abbreviations used in bibliographies are listed in the latest *Mathematical Reviews* annual index. The series abbreviations are also accessible from www.ams.org/publications/. To help in preparing and verifying references, the AMS offers MR Lookup, a Reference Tool for Linking, at www.ams.org/mrlookup/. When the manuscript is submitted, authors should supply the editor with electronic addresses if available. These will be printed after the postal address at the end of the article.

**Electronically prepared manuscripts.** The AMS encourages electronically prepared manuscripts, with a strong preference for $\mathcal{AMS}$-LaTeX. To this end, the Society has prepared $\mathcal{AMS}$-LaTeX author packages for each AMS publication. Author packages include instructions for preparing electronic manuscripts, the *AMS Author Handbook*, samples, and a style file that generates the particular design specifications of that publication series. Though $\mathcal{AMS}$-LaTeX is the highly preferred format of TeX, author packages are also available in $\mathcal{AMS}$-TeX.

Authors may retrieve an author package from e-MATH starting from www.ams.org/tex/ or via FTP to ftp.ams.org (login as anonymous, enter username as password, and type cd pub/author-info). The *AMS Author Handbook* and the *Instruction Manual* are available in PDF format following the author packages link from www.ams.org/tex/. The author package can also be obtained free of charge by sending email to `tech-support@ams.org` (Internet) or from the Publication Division, American Mathematical Society, 201 Charles St., Providence, RI 02904-2294, USA. When requesting an author package, please specify $\mathcal{AMS}$-LaTeX or $\mathcal{AMS}$-TeX and the publication in which your paper will appear. Please be sure to include your complete email address.

**Sending electronic files.** After acceptance, the source file(s) should be sent to the Providence office (this includes any TeX source file, any graphics files, and the DVI or PostScript file).

Before sending the source file, be sure you have proofread your paper carefully. The files you send must be the EXACT files used to generate the proof copy that was accepted for publication. For all publications, authors are required to send a printed copy of their paper, which exactly matches the copy approved for publication, along with any graphics that will appear in the paper.

TeX files may be submitted by email, FTP, or on diskette. The DVI file(s) and PostScript files should be submitted only by FTP or on diskette unless they are encoded properly to submit through email. (DVI files are binary and PostScript files tend to be very large.)

Electronically prepared manuscripts can be sent via email to `pub-submit@ams.org` (Internet). The subject line of the message should include the publication code to identify it as a Memoir. TeX source files, DVI files, and PostScript files can be transferred over the Internet by FTP to the Internet node `e-math.ams.org` (130.44.1.100).

**Electronic graphics.** Comprehensive instructions on preparing graphics are available at `www.ams.org/jourhtml/graphics.html`. A few of the major requirements are given here.

Submit files for graphics as EPS (Encapsulated PostScript) files. This includes graphics originated via a graphics application as well as scanned photographs or other computer-generated images. If this is not possible, TIFF files are acceptable as long as they can be opened in Adobe Photoshop or Illustrator. No matter what method was used to produce the graphic, it is necessary to provide a paper copy to the AMS.

Authors using graphics packages for the creation of electronic art should also avoid the use of any lines thinner than 0.5 points in width. Many graphics packages allow the user to specify a "hairline" for a very thin line. Hairlines often look acceptable when proofed on a typical laser printer. However, when produced on a high-resolution laser imagesetter, hairlines become nearly invisible and will be lost entirely in the final printing process.

Screens should be set to values between 15% and 85%. Screens which fall outside of this range are too light or too dark to print correctly. Variations of screens within a graphic should be no less than 10%.

**Inquiries.** Any inquiries concerning a paper that has been accepted for publication should be sent directly to the Electronic Prepress Department, American Mathematical Society, 201 Charles St., Providence, RI 02904, USA.

# Editors

This journal is designed particularly for long research papers, normally at least 80 pages in length, and groups of cognate papers in pure and applied mathematics. Papers intended for publication in the *Memoirs* should be addressed to one of the following editors. In principle the Memoirs welcomes electronic submissions, and some of the editors, those whose names appear below with an asterisk (*), have indicated that they prefer them. However, editors reserve the right to request hard copies after papers have been submitted electronically. Authors are advised to make preliminary email inquiries to editors about whether they are likely to be able to handle submissions in a particular electronic form.

***Algebra** to ALEXANDER KLESHCHEV, Department of Mathematics, University of Oregon, Eugene, OR 97403-1222; email: ams@noether.uoregon.edu

***Algebra and its application** to MINA TEICHER, Emmy Noether Research Institute for Mathematics, Bar-Ilan University, Ramat-Gan 52900, Israel; email: teicher@macs.biu.ac.il

**Algebraic geometry** to DAN ABRAMOVICH, Department of Mathematics, Brown University, Box 1917, Providence, RI 02912; email: amsedit@math.brown.edu

**Algebraic number theory** to V. KUMAR MURTY, Department of Mathematics, University of Toronto, 100 St. George Street, Toronto, ON M5S 1A1, Canada; email: murty@math.toronto.edu

***Algebraic topology** to ALEJANDRO ADEM, Department of Mathematics, University of British Columbia, Room 121, 1984 Mathematics Road, Vancouver, British Columbia, Canada V6T 1Z2; email: transactions@math.ubc.ca

***Combinatorics** to JOHN R. STEMBRIDGE, Department of Mathematics, University of Michigan, Ann Arbor, Michigan 48109-1109; email: JRS@umich.edu

**Complex analysis and harmonic analysis** to ALEXANDER NAGEL, Department of Mathematics, University of Wisconsin, 480 Lincoln Drive, Madison, WI 53706-1313; email: nagel@math.wisc.edu

***Differential geometry and global analysis** to LISA C. JEFFREY, Department of Mathematics, University of Toronto, 100 St. George St., Toronto, ON Canada M5S 3G3; email: jeffrey@math.toronto.edu

**Dynamical systems and ergodic theory** to AMIE WILKINSON, Department of Mathematics, Northwestern University, 2033 Sheridan Road, Evanston, IL 60208-2730; email: transactions@math.northwestern.edu

***Functional analysis and operator algebras** to MARIUS DADARLAT, Department of Mathematics, Purdue University, 150 N. University St., West Lafayette, IN 47907-2067; email: mdd@math.purdue.edu

***Geometric analysis** to TOBIAS COLDING, Courant Institute, New York University, 251 Mercer St., New York, NY 10012; email: traneditor@cims.nyu.edu

***Geometric topology** to MLADEN BESTVINA, Department of Mathematics, University of Utah, 155 South 1400 East, JWB 233, Salt Lake City, Utah 84112-0090; email: bestvina@math.utah.edu

**Harmonic analysis, representation theory, and Lie theory** to ROBERT J. STANTON, Department of Mathematics, The Ohio State University, 231 West 18th Avenue, Columbus, OH 43210-1174; email: stanton@math.ohio-state.edu

***Logic** to STEFFEN LEMPP, Department of Mathematics, University of Wisconsin, 480 Lincoln Drive, Madison, Wisconsin 53706-1388; email: lempp@math.wisc.edu

***Ordinary differential equations, partial differential equations, and applied mathematics** to PETER W. BATES, Department of Mathematics, Michigan State University, East Lansing, MI 48824-1027; email: bates@math.msu.edu

**Partial differential equations** to GUSTAVO PONCE, Department of Mathematics, South Hall, Room 6607, University of California, Santa Barbara, CA 93106; email: ponce@math.ucsb.edu

***Probability and statistics** to KRZYSZTOF BURDZY, Department of Mathematics, University of Washington, Box 354350, Seattle, Washington 98195-4350; email: burdzy@math.washington.edu

***Real analysis and partial differential equations** to DANIEL TATARU, Department of Mathematics, University of California, Berkeley, Berkeley, CA 94720; email: tataru@math.berkeley.edu

**All other communications to the editors** should be addressed to the Managing Editor, ROBERT GURALNICK, Department of Mathematics, University of Southern California, Los Angeles, CA 90089-1113; email: transams@math.usc.edu

# Titles in This Series

860 **Thomas M. Fiore,** Pseudo limits, biadjoints, and pseudo algebras: Categorical foundations of conformal field theory, 2006

859 **N. Arcozzi, R. Rochberg, and E. Sawyer,** Carleson measures and interpolating sequences for Besov spaces on complex balls, 2006

858 **Enrico Valdinoci, Berardino Sciunzi, and Vasile Ovidiu Savin,** Flat level set regularity of $p$-Laplace phase transitions, 2006

857 **Donatella Danielli, Nicola Garofalo, and Duy-Minh Nhieu,** Non-doubling Ahlfors measures, perimeter measures, and the characterization of the trace spaces of Sobolev functions in Carnot-Carathéodory spaces, 2006

856 **Vladimir Bolotnikov and Harry Dym,** On boundary interpolation for matrix valued Schur functions, 2006

855 **Yevgenia Kashina, Yorck Sommerhäuser, and Yongchang Zhu,** On higher Frobenius-Schur indicators, 2006

854 **Noam Greenberg,** The role of true finiteness in the admissible recursively enumerable degrees, 2006

853 **Joachim Krieger,** Stability of spherically symmetric wave maps, 2006

852 **Viorel Barbu, Irena Lasiecka, and Roberto Triggiani,** Tangential boundary stabilization of Navier-Stokes equations, 2006

851 **Jie Wu,** On maps from loop suspensions to loop spaces and the shuffle relations on the Cohen groups, 2006

850 **Siegfried Echterhoff, S. Kaliszewski, John Quigg, and Iain Raeburn,** A categorical approach to imprimitivity theorems for $C^*$-dynamical systems, 2006

849 **Katsuhiko Kuribayashi, Mamoru Mimura, and Tetsu Nishimoto,** Twisted tensor products related to the cohomology of the classifying spaces of loop groups, 2006

848 **Bob Oliver,** Equivalences of classifying spaces completed at the prime two, 2006

847 **Eric T. Sawyer and Richard L. Wheeden,** Hölder continuity of weak solutions to subelliptic equations with rough coefficients, 2006

846 **Victor Beresnevich, Detta Dickinson, and Sanju Velani,** Measure theoretic laws for lim–sup sets, 2006

845 **Ehud Friedgut, Vojtech Rödl, Andrzej Ruciński, and Prasad V. Tetali,** A Sharp threshold for random graphs with a monochromatic triangle in every edge coloring, 2006

844 **Amadeu Delshams, Rafael de la Llave, and Tere M. Seara,** A geometric mechanism for diffusion in Hamiltonian systems overcoming the large gap problem: Heuristics and rigorous verification on a model, 2006

843 **Denis V. Osin,** Relatively hyperbolic groups: Intrinsic geometry, algebraic properties, and algorithmic problems, 2006

842 **David P. Blecher and Vrej Zarikian,** The calculus of one-sided $M$-ideals and multipliers in operator spaces, 2006

841 **Enrique Artal Bartolo, Pierrette Cassou-Noguès, Ignacio Luengo, and Alejandro Melle Hernández,** Quasi-ordinary power series and their zeta functions, 2005

840 **Sławomir Kołodziej,** The complex Monge-Ampère equation and pluripotential theory, 2005

839 **Mihai Ciucu,** A random tiling model for two dimensional electrostatics, 2005

838 **V. Jurdjevic,** Integrable Hamiltonian systems on complex Lie groups, 2005

837 **Joseph A. Ball and Victor Vinnikov,** Lax-Phillips scattering and conservative linear systems: A Cuntz-algebra multidimensional setting, 2005

836 **H. G. Dales and A. T.-M. Lau,** The second duals of Beurling algbras, 2005

835 **Kiyoshi Igusa,** Higher complex torsion and the framing principle, 2005

834 **Keníchi Ohshika,** Kleinian groups which are limits of geometrically finite groups, 2005

## TITLES IN THIS SERIES

- 833 **Greg Hjorth and Alexander S. Kechris,** Rigidity theorems for actions of product groups and countable Borel equivalence relations, 2005
- 832 **Lee Klingler and Lawrence S. Levy,** Representation type of commutative Noetherian rings III: Global wildness and tameness, 2005
- 831 **K. R. Goodearl and F. Wehrung,** The complete dimension theory of partially ordered systems with equivalence and orthogonality, 2005
- 830 **Jason Fulman, Peter M. Neumann, and Cheryl E. Praeger,** A generating function approach to the enumeration of matrices in classical groups over finite fields, 2005
- 829 **S. G. Bobkov and B. Zegarlinski,** Entropy bounds and isoperimetry, 2005
- 828 **Joel Berman and Paweł M. Idziak,** Generative complexity in algebra, 2005
- 827 **Trevor A. Welsh,** Fermionic expressions for minimal model Virasoro characters, 2005
- 826 **Guy Métivier and Kevin Zumbrun,** Large viscous boundary layers for noncharacteristic nonlinear hyperbolic problems, 2005
- 825 **Yaozhong Hu,** Integral transformations and anticipative calculus for fractional Brownian motions, 2005
- 824 **Luen-Chau Li and Serge Parmentier,** On dynamical Poisson groupoids I, 2005
- 823 **Claus Mokler,** An analogue of a reductive algebraic monoid whose unit group is a Kac-Moody group, 2005
- 822 **Stefano Pigola, Marco Rigoli, and Alberto G. Setti,** Maximum principles on Riemannian manifolds and applications, 2005
- 821 **Nicole Bopp and Hubert Rubenthaler,** Local zeta functions attached to the minimal spherical series for a class of symmetric spaces, 2005
- 820 **Vadim A. Kaimanovich and Mikhail Lyubich,** Conformal and harmonic measures on laminations associated with rational maps, 2005
- 819 **F. Andreatta and E. Z. Goren,** Hilbert modular forms: Mod $p$ and $p$-adic aspects, 2005
- 818 **Tom De Medts,** An algebraic structure for Moufang quadrangles, 2005
- 817 **Javier Fernández de Bobadilla,** Moduli spaces of polynomials in two variables, 2005
- 816 **Francis Clarke,** Necessary conditions in dynamic optimization, 2005
- 815 **Martin Bendersky and Donald M. Davis,** $V_1$-periodic homotopy groups of $SO(n)$, 2004
- 814 **Johannes Huebschmann,** Kähler spaces, nilpotent orbits, and singular reduction, 2004
- 813 **Jeff Groah and Blake Temple,** Shock-wave solutions of the Einstein equations with perfect fluid sources: Existence and consistency by a locally inertial Glimm scheme, 2004
- 812 **Richard D. Canary and Darryl McCullough,** Homotopy equivalences of 3-manifolds and deformation theory of Kleinian groups, 2004
- 811 **Ottmar Loos and Erhard Neher,** Locally finite root systems, 2004
- 810 **W. N. Everitt and L. Markus,** Infinite dimensional complex symplectic spaces, 2004
- 809 **J. T. Cox, D. A. Dawson, and A. Greven,** Mutually catalytic super branching random walks: Large finite systems and renormalization analysis, 2004
- 808 **Hagen Meltzer,** Exceptional vector bundles, tilting sheaves and tilting complexes for weighted projective lines, 2004
- 807 **Carlos A. Cabrelli, Christopher Heil, and Ursula M. Molter,** Self-similarity and multiwavelets in higher dimensions, 2004
- 806 **Spiros A. Argyros and Andreas Tolias,** Methods in the theory of hereditarily indecomposable Banach spaces, 2004

For a complete list of titles in this series, visit the
AMS Bookstore at **www.ams.org/bookstore/**.